T0333979

"This comparative analysis of the evolution of housing markets across Latin American nations fills a much needed gap in the global literature on financialization. Alfonso Valenzuela Aguilera examines a series of case studies, which highlight differing policies and outcomes, and demonstrates the social implications of the variegated financialization of land, real estate, and housing policy across the region. This is a huge contribution to the scholarship and necessary reading for those seeking solutions to the growing problems caused by asset-based policies."

—*Alan Walks, Professor of Urban Geography and Planning, University of Toronto.*

The Financialization of Latin American Real Estate Markets: New Frontiers

The Financialization of Latin American Real Estate Markets: New Frontiers introduces the fundamental principles of urban economics, housing, and large-scale real estate development in Latin America and equips aspiring investors and developers with the foundations for success in a unique, dynamic region. Using case studies from the Americas, this textbook provides a framework for assessing the economic, technological, social, and political forces that shape urban space, helping readers understand the aims and risks of real estate investment. Chapters on economic theory, novel financial instruments, and the regulatory environment connect real-world practice to the latest scholarly conversations in urban planning, real estate finance and development, and regional studies.

Informed by the author's extensive experience as an academic and practitioner throughout the region, this distinctive resource sheds light on the relationship between financial capital and urban form, and places Latin American cities at the center of the urban economy debate.

Features:

- Provides a thorough introduction to the mechanics of real estate markets, grounding spatial and economic theories with practical examples of the tools used to finance urban development in Latin America
- Centers around case studies from Mexico, Brazil, Chile, Panama, Argentina, and Colombia – some of the region's most dynamic markets
- Presents financial instruments such as mortgage-backed securities, collateralized debt obligations, credit default swaps, and real estate investment trusts in a global context
- Examines State policies and programs for housing and infrastructure in Latin America, demonstrating regional patterns and new perspectives
- Covers real estate finance from housing to megaprojects, exploring recent trends in infrastructure, commercial centers, and tourism with an eye toward sustainable financing practices for the future

Suitable for graduate and upper-level undergraduate students of real estate, urban planning, and Latin American studies, *The Financialization of Latin American Real Estate Markets: New Frontiers* also serves as essential reading for professionals in international real estate finance and development.

Alfonso Valenzuela Aguilera is Professor of Urban Planning at the State University of Morelos, and the John Bousfield Distinguished Visitor in Planning at the University of Toronto. He also works as a planning analyst/consultant and advises various citizen boards and councils.

The Financialization of Latin American Real Estate Markets

New Frontiers

Alfonso Valenzuela Aguilera

Routledge
Taylor & Francis Group

NEW YORK AND LONDON

Cover image: © Getty Images/Carlos Silva

First published 2023
by Routledge
605 Third Avenue, New York, NY 10158

and by Routledge
4 Park Square, Milton Park, Abingdon, Oxon, OX14 4RN

Routledge is an imprint of the Taylor & Francis Group, an informa business

Library of Congress Cataloging-in-Publication Data
Names: Valenzuela Aguilera, Alfonso, author.
Title: The financialization of Latin American real estate markets: new frontiers / Alfonso Valenzuela Aguilera.
Description: New York, NY: Routledge, 2022. |
Includes bibliographical references and index.
Identifiers: LCCN 2021061012 (print) | LCCN 2021061013 (ebook) |
ISBN 9780367634872 (paperback) | ISBN 9780367634896 (hardback) |
ISBN 9781003119340 (ebook)
Subjects: LCSH: Real estate investment–Latin America. |
Real property–Latin America. | Financialization–Latin America.
Classification: LCC HD320.5 . V35 2022 (print) |
LCC HD320.5 (ebook) | DDC 332.63/24098–dc23/eng/20220107
LC record available at https://lccn.loc.gov/2021061012
LC ebook record available at https://lccn.loc.gov/2021061013

ISBN: 978-0-367-63489-6 (hbk)
ISBN: 978-0-367-63487-2 (pbk)
ISBN: 978-1-003-11934-0 (ebk)

DOI: 10.1201/9781003119340

Typeset in Goudy
by Newgen Publishing UK

Contents

Preface   ix
Acknowledgments   x

Introduction   1

1 Financial capital and the restructuring of cities   5
The urban economy and the city structure 5
Framing the development of geographical economics 7
Revisiting urban spatial economics 12
The economics of spatial location and the definition of land prices 27
Conclusion: the economic nature of cities 31

2 Theorizing the spatial structure of cities   35
Fictitious capital and real estate markets 35
Revisiting urban land rent theories 39
The formation of urban land prices 42
The three circuits of the spatial economy 49
Social and institutional frameworks 55
Real estate markets in Latin America 58

3 Financial instruments as tools   63
Financial instruments and the location of capital 63
Mortgage-backed securities (MBSs) 67
The impact of collateralized debt obligations (CDOs) 72
Credit default swaps (CDSs) 78
Real estate investment trusts (REITs) 83

4 Financing housing markets   91
Housing: financial markets and the State 91
Chile and progressive housing programs 96
Mexico and private housing corporations 102

Brazil and Minha Casa, Minha Vida *program 113*
Colombia and the production of social housing schemes 119
New perspectives on housing policies in Latin America 125

5   Financing megaprojects in Latin America   130
Megaprojects: finance, markets, and the State 130
Puerto Madero, Buenos Aires 132
Santa Fe, Mexico City 138
Punta Pacífica, Panama City 144
Porto Maravilha, Rio de Janeiro 151

6   Financial instruments and the city   159
International flows of capital and urban governance 159
Financial instruments for infrastructure 163
Financial instruments for shopping centers 176
Financial instruments for tourism 180

**7   Conclusions: sustainable financing practices
of cities**   186
The financialization of living environments 187
Megaprojects: large investments and the State 191
The instrumentalization of capital in the urban realm 194

References   199
Index   216

Preface

Financial capital is shaping cities and restructuring their territories, increasing its dominance among financial actors, markets, and decision-makers. This book provides readers with a basic understanding of the principles that underlie real estate development, urban economic theories, and the particular modes of finance that are being used to reshape Latin American cities, emphasizing the role of land rent theory in the tradition of Marxist analysis as a theoretical foundation, adding to the debate on urban economics and introducing the main concepts pertaining to financial real estate markets.

This cutting-edge volume paves the way for future research, attending to the issue of market dynamics and financial bubbles, which have been cyclical in Latin America both before and in the aftermath of the US house-price collapse that precipitated the global crisis of 2008. This volume introduces readers to the mechanics of real estate markets covering a combination of theories, instruments, and practical examples. Focusing on Latin America, the book reflects the way in which cities follow similar patterns in the design of housing policy and programs, the mechanisms for producing infrastructure in those cities, and the growing disparities that high levels of investment are producing among their citizens.

Acknowledgments

Over the last decade I have many people to thank for their close collaboration, starting with Professor Peter M. Ward at the Lyndon B. Johnson School of Public Affairs, University of Texas at Austin, where I was invited as the Matias Romero Chair in Mexican Studies and organized a Capstone Research Workshop to discuss housing practices and policies. Later, I served as the Mexico Chair at the Faculty of Environmental Design at the University of Calgary, where Professor Saskia Tsenkova not only served as a gracious host but contributed to the discussion of the present research. I would also like to acknowledge the Regional Studies Association and European Union Commission who presented me with the International Award on Local Development for the best paper in 2017, in which the preliminary ideas were developed for the present volume.

A special mention to the University of Toronto for conferring on me the honor of holding the John Bousfield Distinguished Professorship in Planning in the Department of Geography and Planning, where Professor Paul Hess provided invaluable help and the conditions to write most of my research. I should make a special mention of the generous efforts of Alan Walks, who took the time to read a first version of this work and discuss it with me in some depth.

Finally, I would like to extend my gratitude to the Center for Latin America and the Caribbean at the Institute of Social Sciences at the University of Buenos Aires Mexico for hosting me as I wrote the final chapter of the research project; to professors Pedro Pírez and Ivana Socoloff for their questions and acute observations; and to the Universidad Autónoma del Estado de Morelos for providing continuous support throughout the project. A special mention to the editorial team at Routledge for their relentless support along the way, and to my research assistant, María Fernanda Rivero, and Guillermo Romero Tecua serving as a data consultant, both of whom helped me to complete this project.

Introduction

Real estate has become a central object of financialization and its share of the overall stock of capital and of income has grown exponentially in the last decades. However, this debt-based accumulation model, which supports private consumption and enhances real estate markets through financial assets, has led to major crashes that resulted in greater inequality and income disparities in the aftermath of the crises. Even if the system has been somehow resilient during the long-term cycles of capitalism, the periods of instability have furthered income and class divisions as well as ousting the lower-income population from their former territories (Walks, 2010; Aalbers, 2016).

The present volume discusses the role of real estate within the financialization process, the capabilities of the State in regulating and enabling financial markets, and the different mechanisms and instruments that are in place to channel the flows of global capital into real estate markets, becoming one of their major stores of value. The Latin American case entails special features regarding the institutional setup as well as the financial framework, which continue to flourish with heavy State support through favorable policies, subsidies, and other incentives. Along with State endorsement, international banking institutions such as the World Bank have provided substantial loans to enhance the financial and real estate markets, for which the present state of affairs is the result of several circumstances.

Latin American cities are experiencing major spatial transformations fueled by mortgage-backed securities, obligations guaranteed by debt, real estate investment trusts, swaps, futures, derivatives, and other financial institutions and instruments. In order to understand the scale and magnitude of these transformations, this book presents a theoretical framework, a historical overview, and an introduction to some basic principles for large urban operations that draw on examples from Mexico, Brazil, Colombia, Panama, Argentina, and Chile. These cases frame the economic, technological, social, and political forces that shape urban development in

DOI: 10.1201/9781003119340-1

the region and will help readers understand the potential and the risks involved in real estate investment.

The book offers an informed overview of urban economic theory as applied to financial real estate markets and capital investment practices. The analysis presented here focuses primarily on the financial mechanisms that leverage debt, mortgages, and equity positions to enhance urban operations within markets, which are grounded in land rent theory in the tradition of Marxist analysis. To this end, the basic economic principles behind urban growth will be assessed, offering an overview of Latin American real estate markets and investments within a global context.

This volume discusses major features and the latest trends within those areas of finance that are concerned with public real estate markets and presents a series of regional case studies over seven chapters: (1) Financial capital and the restructuring of cities; (2) The real estate market; (3) Financial instruments; (4) Financing the real estate market; (5) Devices of globalization; (6) Alternatives to the financial deregulation of the early twenty-first century; and (7) Conclusions: financing the urban economy.

The first chapter provides an overview of urban economics and the configuration of urban structure, while also examining the financial mechanisms behind investment projects, and the spatial location of investments. The relevance of the principal spatial location models is discussed and assessed for the twenty-first-century metropolis, as are the various mechanisms associated with what is called financialization, that permeate real estate markets and produces a variable geometry of spatial structures in cities. The second chapter examines various interpretations of the urban economic theories in which the land rent theory plays a major role. First introduced by Marx and later modified by several authors, this theory describes the setting of land prices in rural and urban areas, each of which have their own distinct real estate markets. The creation of land prices and the role of fictitious capital in real estate markets are examined, as are the mechanisms for collecting capital gains, enabling an assessment of the real estate financial system and its instruments in Latin America.

The new financial economy is having a major impact in the configuration of cities, reconfiguring the nodes of the global financial network of capital as well as strengthening its structure. The third chapter assesses the new instruments used to finance urban development and attract capital, which include mortgage-backed securities, obligations guaranteed by debt (CDO/CLO), swaps, futures, and derivatives in the real estate market, as well as real estate investment trusts (REITs).

The book then turns to some of the effects of financialization in Latin America. The fourth chapter is an examination of the ways in which major housing programs in Mexico, Brazil, Colombia, and Chile were designed to meet important social challenges, producing millions of housing units, and yet, after several financial crises, resulted in vacant housing stock. It then turns to the impact of financial instruments in Latin America on the housing industry in the last two decades, showing how it drove the over-production of units and later led to the imposition of restrictions on the finance system that provide a stark warning as to the limits of such strategies.

The fifth chapter explores the latest mechanisms affecting the urban economy, including megaprojects and large-scale urban operations that are being developed as investment nodes of the global financial network. From regional infrastructure to shopping centers, business improvement districts, and tourism venues, financial instruments have been designed for the development of large operations that generate liquidity in a real estate market once characterized by a lack of flexibility. Various examples in Latin America show how real estate markets are being reconfigured to enable the accelerated circulation of capital, which is also being enticed by money laundering, resulting in the expansion of the urban construction sector.

The sixth chapter explores international flows of capital and the way they affect urban governance, as large infrastructure works are carried out with major public and private investments, generating mechanisms that lead to appreciating land values and with it, shifts in the economic and territorial balance. First, financial instruments to build public infrastructure are examined, starting with the State-led investment programs of the twentieth century, which later turned into financial instruments that use large public and private capital to fund them. The chapter then offers an overview of shopping centers, malls, and commercial strips as part of a series of emerging urbanization mechanisms affecting land values, followed by a discussion of tourist resorts, which have transformed local and national economies in the region, with a significant impact on the natural environment. Finally, general considerations of equity and other questions are developed from the chapter's discussions of the expanding frontiers of business districts, commercial areas, and tourist enclaves in Latin America.

The seventh chapter states that the urban economy is operating under a different set of rules, which enable financial instruments to serve as mechanisms for expanding real estate markets to give investors the capacity to relocate their assets more quickly and more easily than ever before. It also examines alternatives to securitization-oriented policies and instruments. Land value-capture instruments for financing infrastructure are assessed as possible public policy options, as well as inclusionary housing ordinances that may facilitate comprehensive urban operations. Lastly, reinventing financial instruments by exploring the legal dimensions of real estate development may be the only alternative left for the social reconstruction of communities. Land and real estate value is no longer attached to physical assets, while capital flows, fostered by the different financial mechanisms, and is shaping the cities of the twenty-first century in Latin America.

1 Financial capital and the restructuring of cities

The urban economy and the city structure

According to Brazilian geographer Milton Santos, the structure of urban space responds to shifting forces that create different patterns and constantly readjust its configuration. The components of this form of spatial dialectics assemble in distinct combinations that may produce instability, the specificities of which are determined by historical variations (Santos, 1977, p. 47). Moreover, Santos suggested that two spatial economic circuits combine in developing countries: *traditional* forces, in which consumption is spread and localized to the territory; and *modernizing* forces, which involve technology, capital, and management, and create discontinuous, polarized, and unstable spaces.

While the relationship between space and the economy focusing on patterns, agglomerations, and participating forces has been under assessment for the past two centuries, a new factor has been introduced to spatial configurations, which is the financialization of the real estate market. This factor has compromised the performance of urban planning instruments as the growing scale of capital investments persuades local government officials to endorse large-scale operations, regardless of the environmental, spatial, and social impact they may have in the future.

The financial instruments involved in securitization are highly disruptive of the urban configuration as they enable large international investment funds, hedge funds, pension funds, and insurance companies to hold a stake in real estate assets via the following financial instruments: collateralized debt obligations (CDOs); real estate investment trusts (REITs); asset-backed securities (ABSs); bespoke tranche opportunities (BTOs); collateralized debt positions (CDPs); commercial mortgage-backed securities (CMBSs); and derivatives, all of which enable participation in the growth and operation of major urban properties and infrastructure.

Real estate markets respond to economic cycles, as land is one of the three major factors of production, along with labor and capital. However, the mechanisms that produce urban space do not necessarily respond to classical/neoclassical economic axioms, because cities are also shaped by

DOI: 10.1201/9781003119340-2

public policy rules and regulations, by corporate strategies applied by real estate developers, and by the impact of global capital investments on the urban fabric. As Harvey suggests, changes in capitalist forms of organization, such as the rise of finance capital, multinational corporations, and branch plant manufacturing, allow major investors greater control over progressively larger urban areas, forces that "tend to undermine any structured coherence within a territory" (2001, p. 329). This has happened in cities all over the world, producing, among other things, large vacant housing developments in Mexico City, gentrified historic districts in Barcelona, vacant city sectors in Ordos, land grabbing in the Amazon rainforest, and massive home foreclosures in Oakland, California.

It is in the context of the complexity of the real estate market that we will analyze the impact of the factors of scale, agglomeration, competitiveness, and land use organization. We aim to outline the elements, beyond the apparent efficiency of the market, that intervene in the formation of urban space, revealing the most important components of the spatial economy. Among said components, transportation and communication costs have always had an impact on land prices, limiting access to education, cultural and recreational facilities, and job markets, all of which are highly valued by citizens. Another factor to consider is the *oligopolistic* and *oligopsonistic* dynamics that shape the real estate market, in which major developers are in a position to influence both the supply of and the demand for real estate. This is why urban land rent correlates to construction activity, land use restrictions, intensity of land use, and even land value capture instruments that may be in place in certain areas of the city.

An initial premise in the discussion is that there is a correlation between land prices and building densities, which follow predictable patterns produced by market forces that may determine the spatial distribution of the city (Bertaud, 2015). Consequently, the urban structure will be shaped by the real estate market, which, in recent decades, has integrated financial components that shape the economic principles of supply and demand. By understanding spatial economic models, planners could better predict both localization patterns and densities for the structuring of the urban landscape. Financial capital has unsettled the regular market dynamics that used to respond to the organic growth of cities and, instead, spatial transformations have become a consequence of financial flows rather than real needs. This has resulted in the production of the above-mentioned vacant housing developments as well as the deepening of the mechanisms of socio-spatial segregation and the continued construction of facilities that will be of no further use.

Urban land is considered a commodity as its value can appreciate over time, in the process fueling the corresponding speculation mechanisms, while buildings tend to depreciate. This concept of *commodity* has changed over time, referring to goods that have value and utility, require a low level of processing, and that can be traded without qualitative differentiation as basic components of more complex products. Commodities are traded in spot or futures markets, while, recently, the concept has been extended to any consumer good, which allows the inclusion of financial assets such as currencies, stock indexes, interest, or benchmark rates.

While real estate markets depend on external factors such as credit, interest rates, income, inflation, or consumer preferences, both land and buildings are considered durable consumer goods as well as investment commodities that increase in value over time. These goods are exchanged on the real estate market. However, they are not simply commodities: there is a constitutional right to adequate housing in many Latin American countries and the right to the city has also been granted in the local charters of some major cities.

Framing the development of geographical economics

The analysis of economic models is central to understanding the spatial structure of cities and may be the only way in which public policy can improve the mechanisms that not only govern the formal economy but also have a direct impact on the informal sector. The contemporary city structure is grounded in the industrial era, during which new connections among the territorial units of cities were activated, thereby transforming the social life of their citizens. In this sense, not only can the city be seen as the mechanism of realizing practices that enhance a particular type of society (formerly industrial and now technological) but can also be considered as an outcome in itself (Saunders, 1986, p. 145).

Along the aforementioned lines, the city can be defined as a concentrator of endeavors, needs, and opportunities, due to the interdependent relationships between activities and their spatial correlations. Weber (1958 [1929], p. 72) suggests that cities specialize according to their economic functions, in which certain dynamics attract a large base of consumers, producers, merchants, etc. that will later develop into financial, business, and service centers. Moses and Williamson (1967) argue that the urban structure of cities in the nineteenth century was shaped by the need to locate urban centers near transport terminals to reduce the high costs of the intra-urban transportation of goods, while the introduction of trailer trucks in the twentieth century minimized expenditures by facilitating decentralized transportation schemes. Along these same lines, Hoover and Vernon (1959) suggest that proximity among companies associated with certain products enhances their ability to better adapt to changes in their respective industries, either due to shared technological advances or to the reduction in transportation costs.

The fascination around cities lies in their capacity to foster economic and social transformation via large-scale production that allows for the opening of specialized markets for different products and services, and, as a consequence, urban concentrations come to represent the great stage for the activities of modern life. As Simmel suggests, the city enables producers and consumers to establish interdependent relationships to maximize the benefits obtained from the surpluses mediated by the territory:

> It is rather in transcending this purely tangible extensiveness that the metropolis also becomes the seat of cosmopolitanism. Comparable with the form of the development of wealth (beyond a certain point property increases in ever more rapid progression as out of its inner being) the individual's horizon is enlarged.
>
> (Simmel, 1903, p. 17)

Simmel (1903) argued that the concentration of the population in cities is derived from their great heterogeneity and differentiation between individuals, which, in turn, produces spatial segregation and the segmentation of social relations. Thus, professional specialization stems from the size of the labor market and the spread of productive activities that require operational efficiency. The relevance of the differentiation and *territorialization* produced by increased population density yields a more complex social structure where competition for space increases, as the zoning rationality is governed by real estate market mechanisms that value location, density, aesthetics, and prestige, among other factors. As Simmel warns on urban intervention projects in the city, "the unearned increment of ground rent through a mere increase in traffic brings to the owner profits which are self-generating" (Simmel, 1903, p. 17). It should be again noted that every action that takes place in the urban realm has an economic impact that benefits specific actors. Concentrating specialized functions in the city creates differentiated functional hubs, which are in part defined by their internal consistency. According to Wirth (1988 [1938], p. 177), this dynamic creates physical proximity and simultaneously recreates the social distance derived from competition between individuals.

The discussion of the use of urban development as an instrument for the valorization of capital has long featured in the analytical tradition that dates back to critiques of the industrial capitalist mode of production, among which seminal works by Marx (1998), Alonso (1964), Topalov (1974), Lipietz (1985), Jaramillo González (2009), and Lefebvre (1976) stand out. Since the end of the last century, however, new economic mechanisms, based on the instruments of financialization that enable large-scale international capital investments, have transformed the territorial structure of cities in the short term.

Drawing on Marx (1978, p. 137), we can consider the city instrumental in the consolidation of capital, as it concentrates the resources that enable the hegemonic control of space, even though the urban realm also hosts the revolutionary potential of the proletariat. In Marx's *Grundrisse* (Marx, 2007), space appears as a decisive factor for the circulation of products and the expansion of markets, although it can also become a limiting factor when it increases the cost of the circulation and transport of goods. For Marx, the city is a concentration of both the means of production and the labor force, ensuring maximum productive and distributive efficiency, making it the perfect locus for the capitalist mode of production (Marx, 1971, p. 36). Under this logic, capitalist production based on the maximization of profit subjects working conditions to great strain, which may also extend to the housing situation and family dynamics of the labor force, in the process recreating previous structures that propagated class exploitation (Marx, 1973, p. 564).

As a consequence, the underlying logic of space does not respond to human needs but those of capital, wherein, by controlling the means of production, the dominant political and economic groups also dictate the production of space, which is a major act of territorial misappropriation. As Lefebvre (1976, p. 148) suggests, space has been subverted by the logic of capital into a commodity created by the economic system, in which the relations of both production and consumption intervene, and in which the production of the built environment adopts an economic, and even financial, profit-seeking rationality.

For Marx (1976, p. 140), this creates a divided space that is produced under capitalist rationality characterized by a standard production system that keeps space homogeneous, yet fragmented, due to its commercial exchange value. The urban planning instruments of the twentieth century were converted into ideological weapons to impose the supremacy of central power groups over the peripheral ones (Topalov, 1990, p. 176). This vindicates urban interventions under the dogma of economic development, through which the State can validate, politically and ideologically, the undertaking of public works.

The city, then, is organized through a network of interdependent spatial economies and diseconomies, the agents of which may even operate outside the market (Rémy, 1987). Space thus becomes an object of consumption (merchandise), as with tourism or entertainment, becoming an instrument for the valuation of global capital (Lefebvre, 1973, p. 172). The State endorses the dominance of major corporations through the systematization, homogenization, and disaggregation of urban space. In this sense, the real estate market is closely linked to consumer spending and is directly bound to the general performance of the economy, in which access to mortgages and credit is paramount for the activation of such a market.

It is not difficult to see this in light of recurrent global economic crises – notably during the present 2020 pandemic – as rising unemployment, low interest rates, demographic changes, location, or the excessive expectations of property appreciation all could be tied to fluctuations in the demand for and supply of real estate. These changes depended on (a) variations in production costs due to innovations in new materials, technologies, and construction systems, as well as improvements in transportation; (b) rules and regulations affecting land use, land-use intensity tax, permits, and the recovery of capital gains; (c) institutional provisions that expand access to financing, land reserves, or subsidies grounded in specific legislation and institutional partnerships; (d) credit that may have resulted from negotiations with the Inter-American Development Bank or the World Bank, etc.; and (e) unconventional sources, such as money laundering, based on which significant levels of investment are made. In this sense, financialization has driven an artificial demand for real estate that has altered market mechanisms. It should be emphasized that the sector has become a channel for legitimizing funds obtained from illicit activities that already represent a significant proportion of the market, causing fluctuations that often result in variations in both the equilibrium price and available building stock.

According to Fourquet and Murard (1978, p. 36), the city is the expression of the underlying economic system and reflects a social structure in which space is organized according to production and income distribution, requiring the State and society to solve the economic paradoxes arising between actors. In this framework, cities are considered nodal elements within the urban economy and are established as privileged territories where directive, strategic, and organizational functions take place, all of which are instrumental for the circulation of goods and merchandise and, thus, the exercise and expansion of territorial control. Urban agglomerations enable functional specialization and economies of scale, which reduce transport and distribution costs, thus bringing about greater efficiency in the market as a whole. Although the concentration of goods and services has indisputable advantages, territorial dispersion due to economic or social conditions may prevent those living on the periphery of the city from taking advantage of these centralities. The city concentrates economic activities and functions and, for the central thesis of the present work, the financial rationale dominates the real estate markets.

As various authors have claimed, financial mechanisms are transforming the formerly productive economy by escalating the volume of investment, thereby multiplying its impact on territories (Harvey, 2007; De Mattos, 2007). This situation is key to understanding the transformative dynamics of cities, in which financial assets converge with hedge funds, pension funds, insurance companies, and even money laundering to generate massive capital investment in real estate development projects, infrastructure enterprises, and/or large-scale land acquisitions. These capital flows may only be understood within the framework of a global financial environment where speculative mechanisms capture the future value of assets, even betting against them to benefit from contractions in their performance in the short term.

The real estate market is notable for its capacity to absorb surplus capital derived from over-accumulation processes, thereby establishing a second circuit of capital accumulation, which was predicted to become the dominant track once occupied by industrial production activities (Lefebvre, 1976, p. 46). Therefore, the flow of this financial capital is invested in large housing and infrastructure projects (social housing, facilities, education, prisons, and infrastructure), which were formerly undertaken by the public sector and which, in the process, enable the commodification of goods and services previously considered public. Major infrastructure projects, such as roads and highways, airports, thermoelectric plants, railway systems, residential developments, and extractive industries, have become attractive investment opportunities for multinationals and corporations, which may not necessarily operate in the interests of the common good. Enabling these mechanisms for the privatization of public assets usually requires the deregulation of the financial system through the revision of official norms and bidding mechanisms for public enterprises. Moreover, large infrastructure projects are usually associated with foreign debt flows emanating from international banks and funds.

The privatization of major public works usually results in the loss of sovereignty over the nation's assets, resources, productive activities, and strategic decisions, despite financialization accelerating capital turnover time and representing a *spatial fix* of the over-accumulation paradox, channeling resources through trading instruments with ample flexibility and liquidity. Transport and communications accelerate capital flows by facilitating connectivity between certain areas of the city and other activities necessary for social reproduction (work, health, education, etc.), while functional communication networks enable remote work.

Under the above-described rationale, rights to the future value of assets are purchased, introducing highly volatile risk factors that can enable immediate capitalization but entail obligations that could multiply exponentially at any given time, thus transforming the economic rationality of the real estate market. The privatization of large-scale public works may alter the internal structural coherence of the city, challenging the principle that the urban economy must privilege inhabitants over profit, despite such projects being considered spatial fixes for over-accumulation. Before addressing this issue further, it is worth revisiting the classic spatial economic models in order to develop a common grammar for discussing the economic structure of cities.

Revisiting urban spatial economics

The theories of urban spatial economics are designed to integrate a certain degree of dynamism in their models by taking into account the dialectics of geographic specialization, the intensity of use, and land value. There are limitations to these theories, as only some consider a heterogeneous population, a commercial-industrial economic base, private land ownership, free competition in the market, and the presence of defined urban centers. Here, we will concentrate on the spatial aspects of those theories explaining where economic activity takes place and why, as well as the role played by the spatial concentration of people and activities in urban economies. According to Fujita et al. (1999, p. 9), centripetal forces uphold spatial concentrations of economic activity while alternating with centrifugal impulses that oppose such agglomerations. The analysis undertaken of these models aims to detect the rationalities behind spatial configurations, and even if by no means exhaustive, it may provide a basis for understanding the impact of financial capital on the structure of contemporary cities.

According to the theory of industrial location (Weber, 1958 [1929]), three main factors ought to be taken into account: transport costs, labor expenditures, and agglomeration economies. Location is a key concept for urban economics as the greater the number of skilled workers and facilities, the larger the market for modern trade goods and services. However, these factors may differ from the conditions that are important for understanding residential, commercial, or mixed land use, where purchasing power largely determines the available options and preferences for consumers. A second point to consider is how spatial competition affects different economic, productive, or residential users in their efforts to secure the most advantageous locations in the city. For instance, for housing and lodging, proximity to restaurants, museums, and tourist attractions may be an important factor in the appreciation of land value, while accessibility could trigger the reappraisal of historic areas, generating the effects of gentrification.

Although central locations are not always desirable for high-end residential users or adequate for productive activities that require large surface areas for their production processes, core locations in Latin American cities guarantee accessibility to workplaces, facilities, and services, despite the tradeoff of losing living space. However, social housing developments in the periphery of the city are less accessible than central areas, and the combination of rent/mortgage costs, the lack of amenities, and transportation costs may exceed the point of equilibrium, creating, in some cases, massive vacancy rates in these developments. Further on, we will revisit the main paradigms of urban economics to identify their spatial rationality and present a critical view of the theoretical toolboxes developed to explain urban city size and configuration. Throughout the chapter, we will discuss the key properties of the various models that provide insights into the basic elements of location patterns in the context of the complexity of real urban systems.

The location theory paradigm

According to Fujita et al. (1999, p. 15), economists do not usually address the question of how the economy organizes the use of space and, as a last resort, may turn to Johann Heinrich Von Thünen's model of agricultural land use, known as *location theory* (1826/1966). This model became one of the most influential economic paradigms in history and is based on the assumption that competition among agricultural workers for access to the market resulted in a gradient of land rent values from the center of the town outwards, corresponding to the tradeoff generated by rising transportation costs. This theory suggests that urban concentrations act as optimal configurations for manufacturing, as they guarantee that both the value of products and transportation costs would remain constant given the unlimited demand for goods and services in those centralities, despite the tradeoff of restrictions on housing dimensions or property ownership. The model envisioned a single market surrounded by flat agricultural land where farmers would produce and sell crops at the highest market value, thus yielding maximum net profit.

Von Thünen's model for agricultural location was built upon the concepts of land rent and land use equilibrium, and is considered a seminal work for the economic analysis of urban space, with its analysis of transport costs as related to land price and use considered fundamental for economic modeling. Later, Alonso (1964) built upon Von Thünen's theory to account for interurban variations in land use, in which each use type would have its own rent gradient or bid rent curve, meaning that the maximum amount of rent for a specific location would depend on the kind of land use. Therefore, each land use type will compete for the best location within its individual bid rent curve. Because poor households need to be located near the city center (in order to reduce transportation costs required to access job markets), Alonso argues, they have to compete with commercial and industrial land uses, creating a segregated land-use system that affects the disadvantaged population. One aspect of this, the industrial sector, was described at the turn of the twentieth century by Alfred Weber (1958 [1929]), who formulated a theory of industrial location that modeled the optimal location for a market, one which formed a triangle with two raw material sources and enabled the estimation of the transportation least-cost for bringing the product to market and, later, the cheapest labor cost available.

Another approach to spatial location theories was developed by Walter Christaller (1933), in his *central places theory*, which models how the interaction among concentrations, economies of scale, and transportation enables the market to establish a maximum range that a consumer would travel to access a product and the minimum relative radius between the center and a location validating the offering of a certain good. Christaller suggested

> the terminology is not so important when one is not dealing with abstract theoretical research. It will suffice for us to keep in mind the fact that there are economic laws, which determine the life of the economy, and that there are also, consequently, special economic-geographical laws, such as those which determine the sizes, distribution, and number of towns.
>
> (Christaller, 1966 [1933], pp. 3–4)

Christaller described the emergence of the concept of *central places systems* from a geographical condition that furnished them with a specific *hierarchy* in which central places maintained a market area comprising several subsidiary subcenters, thus confining the urban system. These spatial configurations seek equilibrium, inherent to the most efficient locations, from the interaction between the market and consumers, with the city considered as a hub of centralized functions that respond to its hinterland. Although this theory explains the location of individual products, it is not entirely comprehensive in explaining how equilibrium is maintained in the context of a multiplicity of commodities. Christaller's model solely defines residential location using the distance to the central district, while considering a homogeneous and unvarying space, in contrast with other spatial relationships. In this sense, Christaller not only focused on the location of isolated economic activities but also considered the correlation between different economic activities as well as their position in space, ideas that were later introduced to English-speaking academia by Ullman (1941).

Christaller also examined how the processes by which goods and services are produced intervene in the formation of an urban hierarchy, where a specific good can be sold according to a maximum distance range, using the calculation of the minimum quantity that must be produced of the good to be efficient, as "every good is produced only if its scope exceeds the minimum territorial threshold and is placed on a hierarchical scale of goods represented by the dimension of the respective thresholds" (Camagni, 2004, p. 99). According to this market principle, the location of a center would optimize its organization along with a communications principle that drives transport systems to reach larger centers, while another administrative principle guarantees that the market area of the

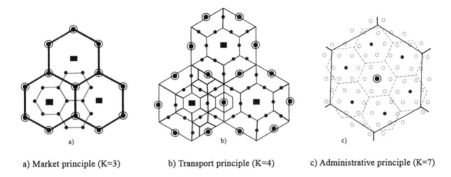

a) Market principle (K=3)   b) Transport principle (K=4)   c) Administrative principle (K=7)

Fig. 1.1 Christaller's central place theory diagrams.

smaller centers is restricted from exceeding the area of the next largest centers (Fig. 1.1).

Building on Christaller's theory, August Lösch proposed a more elaborate scheme that organized the market into a network of hexagonal structures around a central location called the metropolis. The location system implied by his model supported both companies and consumers, calculating the accumulated household demand for space to estimate the entire market area required by a business sector (Mulligan et al., 2012, p. 409). To this end, the different benefits attributed to agglomerations or geographical clustering must be highlighted, such as the concentration of demand for goods and services associated with productive activities, labor pooling, functional specialization through productive chains, and multipurpose and comparative shopping, primarily in the retail and service industries (Mulligan et al., 2012, p. 412). According to Lösch, to maximize profits, each location should excel in its unique type of market, creating a spatial equilibrium when generating both the highest individual profits and the maximum number of independent economic units. The hexagonal lattice configuration proposed by Lösch was meant to be efficient and hierarchical and an attempt to describe, rather than define, the economic and spatial structure of the city.

The concentric ring model

In the early twentieth century, the Chicago School of Urban Ecology provided a pool of ideas that underpinned a broader theoretical model of the contemporary city. Beyond the particularities of that period of history, seminal works of the Chicago School examined urban centralities as the structure that enabled both the creation of mechanisms for community development and generated the spatial patterns that characterized urban life at the time. These works theorized the capitalist urbanization process in the light of profound socioeconomic changes linked to industrial city dynamics.

The Chicago School's research agenda included the compendium of works *The City: Suggestions for Investigation of Human Behavior in the Urban Environment* (Park et al., 1925) and concluded with *Urbanism as a Way of Life* (Wirth, 1988 [1938]), which focused on the spatial impact of immigration and its integration into North American culture: "The slums are also crowded to overflowing with immigrant colonies – the Ghetto, Little Sicily, Greektown, Chinatown – fascinatingly combining old world heritages and American adaptations" (Park et al., 2019, p. 56).

Similar to what is occurring during the contemporary era of technological progress, in the early twentieth century a process of economic reorientation of production took place in western cities, transforming both the economic geography of capital and the instrumental dimension of work. The Chicago School relied on ecological analogies for its analysis of the transformation of social structures and community life, using biological concepts such as *balance, competition, domain,* and *succession* to explain key functions of the city (industry, commerce, and finance, etc.). These functions were situated in strategic locations within the urban structure where processes of *succession* took place whenever a certain sector relocated, leaving the location for another sector to settle.

Although certain inconsistencies in the use of biological categories have subsequently been detected, it represented, for the first time, an attempt by social ecologists to assemble a theoretical apparatus to explain the spatial mechanisms behind the socio-territorial transformation of the city. This process proved to be complicated, as the diversification of occupations created divisions in the social structure – as Burgess suggested:

> These figures also afford some intimation of the complexity and complication of the modern industrial mechanism and the intricate segregation and isolation of divergent economic groups. Interrelated with this economic division of labor there is a corresponding division into social classes and cultural and recreational groups.
>
> (Park et al., 2019, p. 57).

In other words, building on the fundamental economic principle that competition tends toward equilibrium, it was assumed that once the latter was attained, cooperation would come into play.

Following on from this, the physical structuring of urban space was thought to be framed by spatial and temporal relationships that citizens established out of processes of differentiation or adaptation to the environment (McKenzie, 1924, p. 288). In this sense, Burgess examined the mechanisms of urban growth in the period before the great economic depression of 1929, during which physical planning was based on plans, zoning, and regionalization – the most authoritative instruments of public policy. Furthermore, the accelerated expansion of cities involved processes of extension and succession, as well as dynamics of concentration and decentralization, which were claimed to have an impact on social organization (Burgess, 1967, p. 54). Hence, the differentiation derived from the appreciation of land values in certain parts of the city translated into mechanisms for socio-spatial segregation, where the city was fragmented by both zoning provisions and income level, which could displace both land use and population groups.

To this end, the Chicago School framed urban structure as the result of the spatial competition between economic forces that shaped the environment by establishing land-use configurations in the territory. The city's expansion was visualized by Burgess as a socio-territorial process, which he represented with a generic diagram based on the configuration of the city of Chicago in the 1920s, where a series of concentric rings were arrayed outwards from a central business district known as the Loop (Fig. 1.2).

The diagram shown in Fig. 1.2 represents the spatial structure of the city, in which different socio-economic sectors occupied specific rings and, through a succession mechanism, expanded to the point where they settled, outward, in the next concentric ring. These expansion and succession mechanisms are concepts drawn from plant ecology, where opposing but complementary processes of concentration and decentralization take place. For instance, transport systems enabled the concentration of certain land uses, such as suburban housing, while, at the same time, enabling the decentralization of other uses, such as shopping centers in the hinterland.

Accelerated industrial growth in the United States triggered the expansion of cities, leading Burgess to take particular interest in the social impacts of such growth, asking whether there existed a sustainable rate

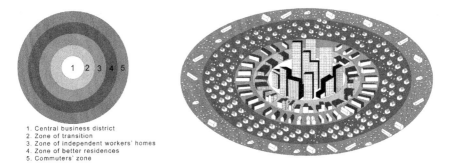

1. Central business district
2. Zone of transition
3. Zone of independent workers' homes
4. Zone of better residences
5. Commuters' zone

Fig. 1.2 Diagram of the concentric zones model.
Source: Burgess, 1925.

of expansion that could keep pace with the controlled changes occurring in social organization (Burgess, 1925, p. 53). He posited that the enigma of urban growth would be solved through the tension between organization and disorganization theories, using the analogy of the processes involved in the phenomenon of *catatonia* occurring in the metabolism of the human body. Burgess considered citizens as an organic part of the city, in which certain urban processes transformed simple elements into complex ones, and energy was also released (Burgess, 1925, p. 53). In this sense, he highlighted sociological aspects, wherein the metaphor of the metabolism of the city implied a moderate degree of disorganization, which resulted in a more adaptive environment. Meanwhile, the theory that accelerated urban expansion pushed disorganization indexes beyond control was developed within the same Chicago School.

While the concentric zone diagram was only intended to describe the processes driving city growth, it also enables the identification of mechanisms of social segregation and spatial fragmentation. At the time that Burgess was writing, slums around the central business district of Chicago were framed as areas of moral decay, even described as purgatories for "lost souls" (Burgess, 1925, p. 56). Social divisions, meanwhile, were considered a manifestation of urban diversity, even becoming a distinctive feature therein, that could potentially broaden diversity by integrating immigrants into the local community:

> This differentiation into natural economic and cultural groupings gives form and character to the city. For segregation offers the group, and thereby the individuals who compose the group, a place and a role in the total organization of city life. Segregation limits development in certain directions, but releases it in others.
>
> (Burgess, 1925, p. 56)

This functional paradox, alongside other contradictions, gave great dynamism to the central area of the city, in terms of the potential to activate the local economy, where "the division of labor in the city likewise illustrates disorganization, reorganization, and increasing differentiation" (Burgess, 1925, p. 56). The city was meant to act as a catalyst in creating a new social order that set citizens apart, less because of its structural advantages but more due to social individualities. However, naturalizing the mechanisms that reproduced the segregation and isolation of disadvantaged economic groups along with this economic division of labor entailed a corresponding division of class and cultural and social affinities (Burgess, 1925, p. 56). The model advanced a concentric configuration of land use that started with the central business district, with locations appraised according to their accessibility to the center.

Based on the foregoing analysis, a rental supply curve can be generated (Fig. 1.3), in which consumers compete to acquire or rent properties

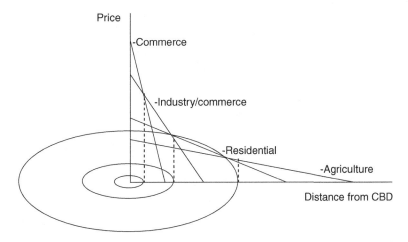

Fig. 1.3 Concentric ring model according to land value.
Source: Burgess, 1925.

closer to the center, thus creating concentric rings of land use. Location patterns were detected based on the expansion and transformation mechanisms observed in the city, wherein the center responds to different uses. Starting from this point, Burgess located the first ring of slums where the immigrant ghettos were to be found:

> clinging close to the skirts of the retail district lies the wholesale and light manufacturing zone. Scattered through this zone and surrounding it, old dilapidated buildings form the homes of the lower working classes, hoboes, and disreputable characters. Here the slums are harbored. Cheap second-hand stores are numerous, and low priced *men only* moving picture and burlesque shows flourish.
>
> (Burgess, 1925)

Considered a working-class transition zone, the second concentric ring of the city housed a manufacturing sector comprising recent migrants, indigents, and criminals, while theories of social disorganization characterized these areas as generators of crime and moral decay. Burgess adds that his research found the greatest concentration of social and environmental deprivation in these areas of physical decay and social disorganization, also stressing, however, the possibility for inhabitants of said areas to transcend these conditions and move out (Burgess, 1925). Second-generation immigrant settlements prevailed in the third ring, in which residents lived near, but not too close, to their workplace, in middle-income housing. This ring was succeeded by a residential ring characterized by a higher income level with better housing (Zone IV), in which lived middle-class native-born Americans, who were transforming single-home communities into apartment house and residential hotel areas. Surrounding the city, the fifth ring housed commuters, providing dormitory suburbs for people who worked in the Loop during the day and came home at night. Burgess later identified two additional zones beyond the built-up area of the city, a sixth ring composed of agricultural districts and lying within the commuter belt, and a seventh corresponding to the hinterland of the metropolis.

The logic behind this model lies in locating the point of greatest economic activity from which land value would appreciate according to accessibility to the center. As Burgess pointed out, "land values, since they reflect movement, afford one of the most sensitive indexes of mobility. The highest land values in Chicago are at the point of greatest mobility in the city, at the corner of State and Madison streets, in the Loop" (Park et al., 2019, p. 61). In the contemporary city, proximity to financial, economic, political, and cultural centers likely increases the value of land prices, in addition to the functional link it creates between these hegemonic spaces and the appraisal of their symbolic power.

According to Evans (1973), the concentric ring model proposed by Burgess to describe the growth of cities suggested that the central part of the city housed the oldest residential properties, whose value was bound to decline, attracting low-income citizens who would settle for inferior living conditions. Therefore, these locations underwent a process of *filtering down* to a more disadvantaged population with fewer resources and a lower standard of living. But this is not universally the case. There are higher-income groups that appreciate central locations and are eager to restore old buildings, even when this leads to the progressive expulsion of former inhabitants while concurrently championing historic districts.

One of the main critiques of Burgess's model is that it naturalized rent as a major factor in the construction of equilibrium in a market-oriented city structure, considering land values the sole indicator of social mobility without taking into account the social, economic, and political implications: "variations in land values, especially where correlated with differences in rents, offer perhaps the best single measure of mobility, and so of all the changes taking place in the expansion and growth of the city" (Park et al., 2019, p. 61). In this sense, the concentric model envisioned the existence of a dominant urban nucleus containing all possible land uses, housing a homogeneous population unaffected by class, race, or gender bias, which, in the long term, resulted in the simplification of natural and organic processes that hid important social and cultural dimensions of city life (Soja, 2000, p. 141).

The radial sector model

Also taking the central business district as a point of departure, land econ-omist Homer Hoyt (1939) drew radial sectors that comprise industrial corridors running along certain avenues of the city adjacent to residential districts notable for their purchasing power. While these sectors could also be located contiguously and did not necessarily comprise a socially diverse population, their location responded to the rationality of the real estate market. This paradigm asserts that land use is distributed in wedged-shaped sectors that run along communication routes and are defined by land values, meaning that, at some point, inhabitants of high-income segments move to better locations, leaving behind more central congested areas in the process of decline, thus establishing the spatial dynamics of the city. Under these circumstances, the disadvantaged population will choose to live in areas with the lowest cost of transportation to their workplaces or, in the case of developing countries, in those areas whose inaccessibility keeps the cost of land so low as to allow self-construction practices or social housing developments (Fig. 1.4).

Given that the radial sector model was a response to specific economic conditions prior to World War II, two and a half decades after the publi-cation of his classic work, Hoyt reflected on the visible changes that cities were undergoing in the postwar period:

> In view of the shifting of uses in the central business districts, the overall decline in the predominance of central retail areas, the rapid growth of office centers in a few cities compared to a static situation in others, the emergence of redeveloped areas, and in-town motels, the former descriptions of patterns in American cities must be revised to conform to the realities of 1964.
>
> (Hoyt, 1964, p. 205)

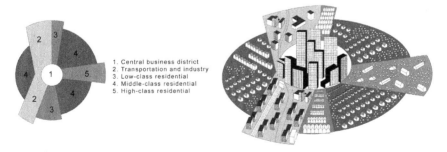

1. Central business district
2. Transportation and industry
3. Low-class residential
4. Middle-class residential
5. High-class residential

Fig. 1.4 Radial sector model diagram.
Source: Hoyt, 1939.

Among the important changes to Hoyt's theory of urban structure in the 1960s was his realization that the principles that guided city growth in the United States were subject to revision when applied to foreign countries. He described the particularities of Latin American cities, where high-income single-family homes and apartments in Guatemala City, Bogota, Lima, La Paz, Quito, Santiago, Buenos Aires, Montevideo, Rio de Janeiro, Sao Paulo, and Caracas were located in areas adjacent to industrial corridors in the city. This revealed socio-spatial fragmentation and income polarization in these cities, as well as the exponential expansion of suburban shopping centers in the context of the far more restricted growth of stores and small businesses in the central districts (Hoyt, 1964, p. 209). In this regard, Hoyt identified ongoing processes in the transformation of territories where

> the tremendous growth of the suburban population, which moved to areas beyond mass transit lines, facilitated by the universal ownership of the automobile, and decline in the numbers and relative incomes of the central city population, invited and made possible this new development in retail shopping.
>
> (Hoyt, 1964, p. 201)

Moreover, he noted the increase of company headquarters in the center of cities such as New York and Washington, DC, where the eviction of slums and run-down areas took place, while the decentralization of main offices to the west and southwest of the original central business districts occurred in the Golden Triangle of Pittsburgh and in the Penn Center of Philadelphia.

The polynuclear model

A third spatial economy model considers a polynuclear configuration, which structures the city around several centralities in which specialized land uses are located at specific focal points (McKenzie, 1933; Harris & Ullman, 1945). These concentrations respond to the requirements of production, thus creating synergy among complementary land uses and reaching consumers who may not be able to access more expensive markets. This built on Christaller but focused not only on the location of isolated economic activities but also considered the correlation between different economic activities and their spatial position. Ullman (1941) introduced these ideas to English-speaking academia, arguing that the combination of these models could yield a better explanation of the urban structure of cities in all its complexity (Fig. 1.5).

Drawing on their influential article on the nature of cities, Harris and Ullman (1945) analyzed the paradoxical character of urban space,

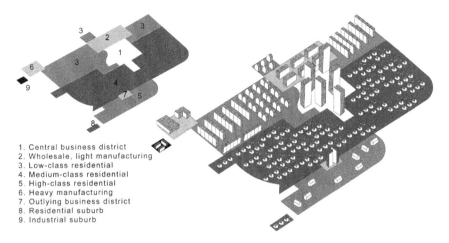

1. Central business district
2. Wholesale, light manufacturing
3. Low-class residential
4. Medium-class residential
5. High-class residential
6. Heavy manufacturing
7. Outlying business district
8. Residential suburb
9. Industrial suburb

Fig. 1.5 Polynuclear model diagram.
Source: Harris & Ullman, 1945.

wherein its heightened growth and economic success are often accompanied by compromising environmental conditions. They claimed that although cities were unique in their particularities, they resembled each other both in terms of operation and location patterns. For instance, while cities in North America tend to develop highly localized resources via tourism, mining, or heavy industry, in Latin American cities, commercial, social, and religious functions are often more important than facilities and services that gravitate around industrial specialization.

To design an effective plan for the improvement or restructuring of a city, Harris and Ullman claimed, planners should begin by analyzing existing land-use patterns to identify the factors that created its configuration, where facilities and services may have played a major role in defining the main activities undertaken within it. This idea was built, in part, on their evaluation of previous spatial models and the suggestion that the forces underlying spatial patterns related to three fundamental components – concentric zones, sectors, and a multiplicity of hubs (see Fig. 1.5). They found that urban growth spread along transportation routes on which similar land uses were found. This assessment identified high-income residential areas in the eastern quadrant of Chicago and low-income housing developments in the southern quadrant, both patterns that were reproduced in a centrifugal direction, namely moving out from the center. However, the authors pointed out that many cities' land-use patterns develop around several *discrete nuclei*, some of which may correspond to the historic districts of the city, while others were generated by the progress of specialization and migration. Their research found that the existence of these separate and differentiated nuclei reflected the combination of four factors:

1   Certain activities required *particular facilities*, such as transport connections, large areas of land, and proximity to the center, etc.;
2   There are *complementary relationships* between different activities or specialized services. In the case of retail, consumers may be attracted by the concentration of specialized stores and would, thus, be able to compare products before buying, while financial districts benefit from rapid communication when located in the same district;
3   The presence of *incompatible uses*, such as industrial activities conducted near high-income residential areas; bars and nightclubs located near schools; or contaminating industries located near natural reserves; and
4   Certain activities would not be able to access prime locations in the city due to the high rent, especially if their space needs are particularly large, such as those required by assembly plants, department stores, parking lots, or social housing.

It is through the combination of these factors that centralities, districts, and regions are linked via interurban transport systems to businesses, wholesale facilities, heavy industry, services, and other specialized activities. However, location patterns may also be related to individual economic capacity, as

high-class districts are likely to be on well-drained, high land, and away from nuisances such as noise, odors, smoke, and railroad lines. Low-class districts are likely to arise near factories and railroad districts, wherever located in the city. Because of the obsolescence of structures, the older inner margins of residential districts are fertile fields for invasion by groups unable to pay high rents.

(Harris & Ullman, 1945, p. 16)

Smaller nuclei can be formed around parks, cultural centers, or universities, forming neighborhoods or small communities. We must highlight the fact that car ownership transformed the urban structure through the creation of the suburbs and the industrial or residential satellite cities that, along with highways, suburban train systems, and electric trams, enabled the expansive suburban growth of cities. The concentric and sectorial theories contend that there was a general tendency for housing in the center of the city to devaluate whenever new construction arose on the outer limits of the city.

Based on an economic analysis grounded on the concentration of capital and commercial and industrial activities, whenever the real estate market depreciates, the amount of vacant, under-occupied, or informally occupied land in the central district increases and could later be replaced by activities generating higher rental value. This points to the high cost of maintaining concentrations of the marginal population in the center of the city and invokes questions of social justice and the right to the city, showing that what is also at stake is the concept of citizenship, where "a separate political status of many suburbs results in a lack of civic responsibility for the problems and expenses of the city in which the suburbanites work" (Harris & Ullman, 1945, p. 17).

Almost two decades after Harris and Ullman's seminal work, the latter reflected on the spatial and financial transformations inherent in the implementation of new mass public transport systems, where suburban locations appreciate in value when made more accessible, thus opening new real estate markets for the higher-income population (Ullman, 1962, p. 12). In this regard, Ullman suggested that urban expansion pointed to new vectors of development: "As urban transport improves, cities not only can expand in the area, but the range of location choice is widened; the most desirable sites within a city can be reached and developed according to their intrinsic advantages" (Ullman, 1962, p. 16). In other words, the level of access to the city center ceases to be the most important consideration in terms of location whenever the location's own attractions come to the fore. Moreover, the reduction of travel time using public transport broadens residential options and extends them to areas of greater scenic beauty or access to more spacious grounds. Ullman thus confirmed his previous theory, which posited that the mere dispersal of housing to the suburbs (so-called *urban sprawl*) did not solve the dilemmas of growth. Dispersal entailed multiple inefficiencies, including the costs of providing infrastructure networks, amenities, and services, in addition to promoting car dependency and consequently increasing air pollution. In sum, the consolidation of a multinuclear structure enables competitive advantages to develop based on a region's economic activity, generating territorial specialization based on cultural, sports, commercial, industrial, and educational hubs.

Ullman did not acknowledge that the social causes of inequity might be a structural element of the capitalist economic system, that is to say, as contributing to the inefficiencies inherent to the functions of the city. He did remark, however,

> As a citizen, I recognize that the major problem of cities – slums and the gray area – cannot be tolerated. We may well have to eliminate them before we eliminate all the causes, including poverty, ignorance, and racial discrimination against the new arrivals, or the other manifold ills of our society both old and new.
>
> (Ullman, 1962, p. 22)

While the city is an economic actor, it cannot renounce its responsibilities to provide social justice and equity, which are instrumental in providing adequate access to the city for all citizens.

Concentration and centralities have been key concepts in urban economics for many years, and in the mid-1990s they were reintroduced into the debate around the pressing issues of development. Porter (1995, p. 55) argued that the lack of jobs and businesses within inner cities fueled a cycle of poverty and other social problems, which demanded social spending around the needs of individuals rather than efforts to create economically viable companies. Although the inner city cannot be isolated from the rest of the surrounding economy, Porter's critique contends that the real challenge is to create wealth rather than distribute it. The latter position, which is inscribed in the neoclassical economic paradigm, ignores the complexities of inner-city areas, where the concentration of all kinds of socioeconomic problems results from redistribution mechanisms, and, ultimately, calls the right to the city into question.

Despite the foregoing, Porter identifies four main advantages of the inner city: strategic location, local market demand, human capital, and integration with regional clusters. Said clusters emerge when complementary activities, such as tech companies and research centers, coincide, thus driving economic development at strategic locations in high-rent areas in which major business centers as well as transportation and communication hubs concentrate, providing such locations a competitive advantage (Porter, 1995, p. 56). In this sense, the concentration of economic activities in the inner city may stimulate local market demands that drive opportunities for city-based entrepreneurs.

Although Porter's model proposes the acceptance of a new paradigm for the inner city based on an economic rather than social perspective – which, incidentally, could have disastrous consequences for low-income inner-city residents – he also suggested that a network of local businesses should be built to create an economic base that would reduce the need for social services that focus on consolidating the local economy. In this sense, financial institutions could play a leading role in supporting both inner-city businesses and community-based organizations that, traditionally, have been instrumental in developing social housing, public spaces, and social programs for local residents.

After revisiting the principal models of urban spatial structure describing the internal social, spatial, and economic structure of cities as heuristic devices that present a basic conceptual framework for spatial analysis, the following section explores the uses of urban economic theory to understand the location of uses in cities.

The economics of spatial location and the definition of land prices

Taking O'Sullivan's five axioms of urban economics as guiding principles for urban economic theory (2012, pp. 7–11), we will assess their relevance in the context of the new financial instruments channeling global capital into contemporary real estate markets.

The first axiom establishes that prices adjust to achieve locational equilibrium, namely, when there is no motivation to change residence or location, the price of real estate will regulate itself accordingly. Conversely, whenever a property presents more attractive features, its value will increase beyond that of a less desirable asset in competitive circumstances. These market mechanisms create residential segregation, as the underprivileged population will become concentrated in the most undesirable parts of the city due to their disadvantageous topography, inaccessibility, or stigma, while the affluent population will be in the position to choose central locations that are well connected, served, and surveilled. Large capital investment in the real estate market can develop thousands of social housing units in a few years; build whole sectors of a city, including shopping complexes, nearby (or even *ghost cities*); or contribute to the overproduction of condominium towers. In these circumstances, prices take longer to achieve equilibrium and, instead, real estate bubbles are generated through speculation, apparent economic prosperity, low interest rates, and easy-to-access credit and mortgages.

The second axiom suggests that one predominant land use can generate a self-reinforcing effect that strengthens the original appeal of the area. Therefore, it is common for certain districts or streets in the city to specialize in commercializing a particular type of product or service, such as restaurants or music, electronics, or appliances stores. When there is a contiguous offer of a product, customers can compare before buying, acquainting themselves with new trends in the market, and even find complementary products to those originally sought. This functional specialization occurs in central business districts, historic centers, shopping centers, museums, entertainment districts, creative industry hubs, industrial parks, *maquiladoras*, tourist complexes, and tech parks. Furthermore, financial instruments may enable the concentration of land use in a short period of time, reducing the opportunity to assimilate said use into the urban fabric.

The third axiom states that externalities produce inefficiencies in the real estate market, as they induce the transfer of these external costs to third parties. This is the rationality behind mechanisms for the recovery of capital gains, which seek to calculate the costs and redistribute the benefits of urban interventions. For instance, when the local government of a city uses public resources to build a park in a certain area of the city, this both generates benefits for the population in general and also contributes to raising the land value of surrounding properties. This positive externality of appreciating land values in a certain area of the city results from an institutional action undertaken with public money, which also creates economic disparities. In such cases, the operation of markets is not socially efficient, for which reason alternative mechanisms to distribute value can be put in place through a system of taxes and incentives that more efficiently recover the costs and redistribute the benefits of public actions.

The fourth axiom establishes that production is subject to economies of scale, which reduce the cost of components, transport, and land, etc. In this sense, the concentration of the population enables the conditions for the mass production of housing, infrastructure, and services. In the case of large cities in the developing world, economies of scale apply to the informal housing sector, where the low-income population takes advantage of economies of scale by accessing low-cost materials to self-construct irregular settlements on land on the periphery of the city, even if that entails living far from the workplace and amenities. While economies of scale apply when large capital investments are channeled through financial instruments, this could have adverse effects in the case of rapid urbanization, as basic services, infrastructure, and amenities may not always follow.

The fifth, and final, axiom suggests that competition tends to reduce economic gains to zero. This entails a spatial dimension, in which companies see their profits affected by the arrival of competitors in their vicinity, as new businesses drive down prices and, thus, profits. For instance, during real estate bubbles, shopping centers are built close to each other, meaning that the businesses they host will be affected first by the urban decline provoked by an economic crisis. The case of the ghost town of Ordos in China is illustrative of this axiom, in that the conditions to create a climate of territorial competition did not exist due to a lack of demand in the city. Another example is the exponential growth of social housing developments on the periphery of Mexican cities, as a result of which construction companies entered the stock market to capitalize on production, creating a housing bubble and distorting the market to the point that it finally collapsed, leaving more than six million abandoned social housing units.

The foregoing axioms show that, in the real estate market, equilibrium depends on different factors that intervene in the appreciation of property. While the value of buildings tends to depreciate, land prices tend to increase over time, although these dynamics are not necessarily permanent and may change due to external factors such as changes in accessibility, services, facilities, infrastructure, or major urban operations that transform entire sectors of the city.

Public policy and institutional mechanisms may also affect the value of property through changes made to density and zoning ordinances. The latter area of public policy involves the application of economic criteria to assign the most profitable land use, with commercial, urban, and residential uses being more profitable than residential, rural, and social housing ones, even when it may not necessarily be the most equitable option. As density restrictions also potentially determine land value, there will be a tendency to seek, via the application of said constrains, the maximum turnover of clientele at any given location, as is the case with commercial avenues or central business districts.

The parameters discussed above are determined by urban regulations intended to protect the spatial equilibrium of the city. To this end, fiscal instruments are introduced as part of a legal and legitimate system for ensuring certain types of development, densities, or combinations of land uses. Incentives can be used in the latter case by conditioning increased building densities or changes to zoning ordinances in exchange for investments in affordable housing, the construction of infrastructure, or the integration of public spaces or cultural facilities into the land use proposed.

It has been argued that the nature of neoclassical economic analysis makes it possible to predict behavior or forecast decisions based on specific conditions related to the allocation of resources within a market. Drawing from the theoretical foundations of the social sciences, there are explanatory and normative approaches to identifying the causes and ideal application of certain financing practices. The disciplines of both economics and urban planning intervene in the operation of real estate markets, generating complex regulatory mechanisms intended to serve the public interest, but which do not always comply with this mandate. In order to understand these mechanisms, we will use tools from both disciplines to demonstrate that the market may seek the maximization of profit and benefit for the actors involved – both individually and collectively. In this sense, the mechanisms for defining land prices give us an insight into the framework in which the urban economy operates, specifically in defining the price of housing units, where the interaction between an inelastic supply and the corresponding demand establishes the equilibrium price when the number of housing units required is equivalent to the quantity supplied.

Figure 1.6 shows that the equilibrium price is found at the point where the demand curve D intersects the supply curve S, setting the price OP, where the quantity OX is supplied and demanded equally. When demand D1 increases, the price rises from OP to OP1, at which point the new equilibrium will be found. Increasing the price per unit will encourage an upturn in supply through the construction of new housing units (Fig. 1.6b), increasing supply from S to S1, which, in turn, produces a reduction in

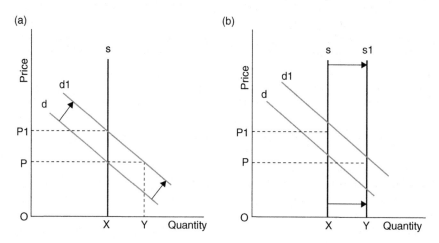

Fig. 1.6 Application of demand-supply analysis to urban accommodation.

price until a level of profit coinciding with OP is reached, namely, by supplying the market with the quantity of units OY.

In the foregoing example, producers respond to the demand of motivated consumers to maximize their profits through market fluctuations. Although the pricing system affects the definition of land use for certain functions, the designation of price considers criteria based on not only economic efficiency but also urban planning, while considerations of equity, redistribution, and social justice may also intervene. Therefore, costs and benefits – also characterized as externalities – are revealed as key factors in estimating the price of real estate properties. For instance, the construction of high-density buildings in the central district of a city is generally associated with an increase in the circulation of motor vehicles, thereby worsening traffic conditions in the area and directly distressing residents, who, ultimately, pay the indirect costs of said construction.

One of the central points of this book is that while the economic mechanisms that structure the city are grounded on the law of supply and demand, in recent decades what this means has changed as the emergence of global financial capital transformed the speed and direction of city development. It has, as noted above, fueled overproduction in the construction industry and led to the development of large vacant housing developments and uninhabited condominium towers, gentrification of historic districts, oversupply of tourism infrastructure, oversaturation of shopping centers, and construction of toll roads and highways that exact high costs for residents.

Moreover, financial capital has demanded the deregulation of the instruments in place to police it. Implicit in these are strong externalities, such as the shortage of basic services and amenities, the expulsion of low-income residents who cannot cope with either the impact of gentrification on their neighborhood or, conversely, the depreciation of real estate properties after housing bubbles burst. On top of this, the public shoulders the cost of maintaining infrastructure and semi-vacant housing developments and the burden generated by the reproduction of mechanisms of spatial segregation that increase socioeconomic polarization in the population. These strains reduce the prospect of moving toward the creation of an equitable, balanced, and just city. Therefore, a proper understanding of these new economic forces is essential knowledge for urban planning professionals to ensure efficient public spending and improve the rationality of public policy decisions. It underlines the need for public intervention in the real estate market, major urban operations, infrastructure projects, and urban regeneration programs.

Conclusion: the economic nature of cities

Urban economic models were created to explain the operation of cities and, thus, prescribe the rational and equitable use of resources. They enable anticipation of the possible impacts of public policy on the city and give market analysts the necessary tools to transition from abstract to operational models. Using this information, 30–35% of international bank loans have traditionally been allocated to urban operations, with these financial institutions claiming that investments in this sector have far-reaching positive impacts, both directly and indirectly, on the general population.

For instance, a project whose main objective is to bring water to a certain area of the city will indirectly increase the value of its land prices, affecting housing and residential demand, which, in turn, may trigger calls for infrastructure or even alter the dynamics of urban transport to and through the area. To understand the impact of these large urban projects on conditions under which a city functions, then, it is necessary to acknowledge that in the last half-century large metropolises have been characterized by accelerated demographic growth, income polarization, and unequal access to technology. This, allied with a relative reduction in the cost of public transport, suggests that the direct and indirect impacts of urban operations must be questioned from different angles. This is also the case for the installation of new modes of transport, such as high-speed trains, that despite reducing commuting time also have a direct impact on the urban structure and the location patterns of industrial districts, residential areas, and shopping centers, where new investments take place. These effects on the urban structure have a territorial impact beyond the mere setting up of the system.

Cities have become instrumental for establishing the basis of economic growth, as they rely on the income elasticity of demand for urban goods and services, which is usually high and increases in line with income (Artle, 1972, p. 61). Mohan argued that, in a broad economic context, the interdependence between the production of goods and services and sectors of ever-greater significance for the city economy such as banking, insurance, and marketing affect the demand for housing and residential locations and other types of infrastructure and, as a consequence, have transformed transportation patterns (1979, p. ix). For this reason, agglomerations profit from economies of scale in terms of the interaction between education and research (which are instrumental for technological development), as well as from cultural and entertainment activities, which require a minimum audience size to be profitable.

One facet of this is the relationship between the city center and its hinterland, where the density, size, and productive complexity of urban concentration have generated the need for specialized facilities and services in specific geographical locations. Future trends in urban planning will be oriented around communications and technology as indispensable components in the production and distribution of goods. This becomes clear when one recognizes that the stability of cities has always been related to their ability to cope with change, in that, as industries have life cycles, the more diversified their economic activities, the more likely a city will be able to adapt to change.

As Jane Jacobs commented in her seminal work (1970), the internal inefficiencies and impracticalities inherent to cities allow them to deal with the uncertainties of technological change and the economic environment. In the case of cities in developing countries, other variables such as income polarization, social segregation, and spatial fragmentation lead to higher-income sectors having closer ties with global cities than with their hinterlands. These differences produce a segmented market divided into formal and informal economies that may be distinguished for analysis, but that, in real terms, exist in a particular economic symbiosis (Valenzuela Aguilera & Tsenkova, 2019). In this sense, while the income range of the informal sector can be extensive, in reality the workforce of small manufacturers, artisans, and retailers represents an economically vulnerable population that travels long hours to their workplaces and is deprived of social benefits. Moreover, the mechanisms of production in cities entail high migration rates, low wages, and little access to capital, all of which favor the establishment of marginal settlements, characterized by irregular property rights and a lack of adequate services, equipment, and transportation, which comprise the largest areas of Latin American cities.

According to Mohan (1979), the high ratios between both capital and land and capital and labor that are found in the city ensure, in industrialized countries, more options for location whenever capital is more accessible for a similar population size, while, in the cities of the developing world, location decisions for the low-income population are limited, to say the least. Finally, although the economic modeling of cities should respond to sound cost-benefit objectives for urban interventions, and despite circumstances when the projected results contrast to those expected, the financialization of the economy has led to another rationality based on revenues and economic performance. Even when the models used for urban interventions have a certain degree of abstraction, specific mechanisms should be made available to anticipate the results that will be obtained in the complex reality represented by said models, especially when they shape subsequent public policy.

Summarizing, city structure is shaped by economic forces through a series of arrangements and variations that may produce discontinuities, polarization, and disequilibrium in the territory. Financial instruments may affect the coherence of the city's spatial configuration due to the scale and magnitude of the urban interventions that global capital can unleash. Analysis of key elements such as densities, land use, transportation costs, connectivity, and other external factors enable the identification of spatial growth patterns. Economic models are key to understanding the spatial structure of cities, which are subject to a specific social organization that considers space a commodity, an object of consumption, or as having both exchange and use-value. Urban planning instruments have also been used as ideological instruments for legitimizing public actions and private interests, meaning that, with the growing dominance of financial, capital investment begins to control the production of space through the incorporation of fictitious capital into the urban fabric, a subject we will introduce in the next chapter. These mechanisms affect the structural coherence of the city as infrastructure will no longer respond to the needs of citizens and will, instead, respond to the future value of financial assets.

Urban economic theories provide a theoretical and general framework that may be useful for envisioning solutions to the urban challenges of geographic specialization, the intensity of land use, and land values. In this context, different dynamics, which involve centripetal forces and centrifugal impulses, shape agglomerations in which spatial rationality may be categorized. The concentric model has been key to understanding the different processes involved in city growth, by means of correlating labor categories to class divisions and identifying adaptation mechanisms. Also, spatial economics identifies processes of concentration and decentralization, turning the city into a catalyst for economic activities, in which densities, land use, and mobility are key factors. However, even when urban ecology models provide a framework for analysis of the structure of cities, they do not address a central issue, which is the identification of the socioeconomic processes that determine the location for different groups and activities within the urban structure. For instance, the radial sector model described the dynamics through which land prices and the cost of transportation play a key role in structuring the city, leading to the growth of business centers in locations other than the traditional city center, wherein retail may either decline or flourish depending on the economic dynamics, and suburban areas develop along mass transit lines. All of these economic features transform the physical configuration of the city and shape its structure.

Location theory provides an understanding of the importance of competition to land rent, where the interaction between the maximization of profit and transportation costs establishes the location of economic activities utilizing specific hierarchies. This theory envisioned hexagonal structures as the ideal configurations in which space could be organized in such a way as to obtain maximum efficiency. However, because this model considers the location of economic activities as a product of the inherent characteristics of physical space, it fails to address the causes of the social division of the urban structure, while the multi-nuclear model, which structures the territory into specialized hubs that interact and create synergies among the nuclei, applies a more complex configuration. To this end, public transport systems enable the expansion of the urban structure and the appreciation of land values beyond the city's centralities, thus widening the range of locations available to the local population, although suburban locations may entail various inefficiencies in the long term.

Spatial location may define land prices, which can be adjusted to achieve locational equilibrium, wherein predominant land use in a location tends to attract similar activities to that specific area of the city. Moreover, externalities should benefit the actors responsible for them, whereas competition should reduce profit to zero whenever monopolistic mechanisms are not introduced or large capital investment does not alter the conditions in which the real estate market operates. Finally, public policy and institutional mechanisms may intervene to protect the public interest and guarantee the spatial equilibrium of the city, regulating the changes to the speed and direction of urban invention that financial capital brings in the real estate market.

2   Theorizing the spatial structure of cities

Fictitious capital and real estate markets

According to Karl Marx, *fictitious capital* refers to financial assets that do not correspond to real capital, since they have no material basis, as either commodities[1] or productive activities. The creation of fictitious capital derives from the anticipation of the process by which financial assets undergo capital valorization, namely, the expectation of a future increase in value. In this sense, Marx claims that the formation of fictitious capital, also known as *capitalization*, consists of the conversion of income or assets into capital out of revenues from financial instruments, such as credit instruments, government bonds, certificates, and shares (Marx & Engels, 2010, p. 597), which enables the holder to participate in financial activities with a fraction of the surplus required. In the case of government bonds, they

> are capital only for the buyer, for whom they represent the purchase price, the capital he invested in them. In themselves, they are not capital, but merely debt claims. If mortgages, they are mere titles on future ground rent. And if they are shares of stock, they are mere titles of ownership, which entitle the holder to a share in future surplus value.
>
> (Marx & Engels, 2010, p. 455)

Harvey argues that the creation of money does not guarantee its conversion into capital, due to the absence of a material basis, such as commodities or productive activities (fixed capital), which can be traded against their inherent surplus value, which has implicit risks:

> The category of *fictitious capital* is in fact, implied whenever credit is extended in advance, in anticipation of future labor as a counter-value. It permits a smooth switch of over-accumulating circulating capital into fixed capital formation – a process that can disguise the appearance of crises entirely in the short run.
>
> (Harvey, 1999, p. 95)

DOI: 10.1201/9781003119340-3

This condition makes financial ventures highly volatile, as they place the credit system, commonly described as a necessary evil of capitalism, at the forefront of the economy. For instance, to overcome the barriers that fixed capital may encounter in circulation, the credit system enables the creation of fictitious capital through financial instruments that solve, at least temporarily, the over-accumulation paradox. While capital may be considered fictitious to the extent that it circulates without any production having been realized, it represents a claim on the future valorization of an asset. However, its speculative nature means it plays an extremely ambivalent role – it may end up advancing capitalist development and capital accumulation beyond the limits of what the economy can withstand, paving the way for a financial crisis.

From a Marxist perspective, fictitious capital loses its capacity to represent potential money-capital during financial crises or periods of economic depression, where its value falls either when interest rates rise or as a result of a shortage of credit. Marx points out that reduced investment in these securities may not correspond to actual capital and, instead, speaks to the solvency of those who own them (Marx & Engels, 2010: 492). In times of crisis, the amount of available fictitious money capital reduces considerably, along with the ability of its owners to use it as collateral to obtain bank credits: "this has the effect of creating a self-sustaining demand for financial products, since the rise in asset prices increases the available collateral and thus frees up loans for buying new securities" (Durand, 2017, p. 69).[2]

In the foregoing scenario, the value of fictitious capital corresponds to that of expected revenues capitalized at the rate of interest, thus relating to the supply of and demand for money capital. Paradoxically, markets for such capital are vital to the survival of capitalism, as they represent the only way to guarantee the continuous flow of interest-bearing assets (Harvey, 1999, p. 278). In this sense, stocks and shares are tradable titles that can be used for the future valorization of the corresponding financial assets and can be traded, converted into money, or shared with other investors, based on their liquidity. As all of these operations are possible within the stock market, the circulation of property rights is guaranteed and will set the conditions for the establishment of financial instruments for use in real estate. Governments can share a portion of tax revenues, fund infrastructure projects, use pension and hedge funds, provide collateral for mortgages, trade commodities, or endorse construction projects without undertaking any further transactions related to the physical condition, location, or stage of construction. Therefore, rights to land, mining, oil extraction, and natural resources can be traded in the form of futures, derivatives, or assets regardless of the consequences that these enterprises may have for the general population. Under these circumstances, land becomes a financial asset or form of interest-bearing capital that is traded according to a claim on future profits from the use of the land (Harvey, 1999, p. 348).

Although Marx did not undertake a detailed analysis of land markets, as it was not considered a product of human labor at the time, land can be assigned a price and be traded as a commodity if its value is capitalized as rent. As has been previously argued, land value is not related to the land itself and depends, instead, on the ground rent it yields, which corresponds to an interest-bearing investment secured via a claim upon anticipated future returns, in the same fashion as fictitious capital (Harvey, 1999, p. 367). This mechanism facilitates a whole range of financial products, such as future rental income and real estate investment trusts, through which land titles can be traded internationally and credit can be used both to leverage capital accumulation and as a tool for speculation and over-accumulation.

The financial innovations of the last three decades have impacted cities as a result of the property market crises with which they are associated, where the credit system fueled surplus capital absorption via the real estate market, employing different variations of financial mechanisms. Marx was particularly visionary in this respect, theorizing that the majority of a financial institution's capital was *purely fictitious*, consisting of claims, government securities, and stocks, and that this capital represented people's savings, whether interest-bearing or not. Moreover, although the capital value of said securities was considered deceptive, the credit system was able to create capital associated with such titles of ownership (Marx & Engels, 2010, p. 464). Harvey (1999, p. 293) considers that capitalism has had to evolve a sophisticated credit system and create fictitious forms of capital to survive within the capital circulation process. Even so, Marx warns of the perils of considering that capital could exist simultaneously as *capital value* via titles of ownership (stocks) and as the *actual capital* invested in the enterprises to which the stock pertains, stating that capital exists only in the latter form, as "a share of a stock is merely a title of ownership to a certain portion of the surplus value to be realized by it" (Marx & Engels, 2010, p. 467). However, the introduction of fictitious capital as a major factor in the financial markets is a defining feature of the economic rationality of contemporary capitalism, part of what is known as financialization, which also encompasses a series of concurrent processes that include the deregulation of financial markets, increased debt across the spectrum, and the sophistication and interdependence of financial instruments.

In the context of real estate markets, for instance, once regulations had been dispensed with, mortgages were found to be useful as building blocks for more complex financial instruments, wherein packed mortgages, known as *NINJA* loans, could be extended even if recipients were unemployed, had no income, or assets as a collateral, and yet, once packed they were sold as securities. Despite the ethical considerations, stockbrokers speculated on these financial products, even betting against the price fluctuations in their own clients' portfolios. Another mechanism of speculation targeted unproductive land that could, nevertheless, fuel a fictitious accumulation process in the event that the titles were to be used as collateral for other sales and purchases (Harvey, 1999, p. 287). Speculation on housing stock had not been a profitable activity until fictitious capital flowed into the housing market, in light of the soaring rate of return on investment. Credit expansion propelled housing prices, as high profits and commissions became common in real estate markets by the late 1990s when financial instruments, such as collateralized debt obligations, served to package subprime credits with solid assets, which were then marketed as prime products worldwide. These procedures later developed into a major financial crisis when multiple traders burst the housing bubble by cashing in their securities simultaneously (Harvey, 2014, p. 32).

This scenario was only made possible by the deregulation of the financial markets and the expansion of predatory practices within the credit system, which triggered speculation in asset values and the practices described by Harvey as *accumulation by dispossession*. These mechanisms created speculative bubbles that ultimately burst, resulting in severe financial and commercial crises in the United States and Latin America (Harvey, 2014, p. 101), although this was not an isolated chapter in the history of financial capital. Dating back to the nineteenth century, Latin America has a long history of infrastructure projects, such as large railway systems, established through financial mechanisms, creating property market booms and speculative activities from which bankers and financial agents profited, obtaining fictitious capital for short-term gains. The subsequent failure of long-term investment "led Marx to speak of the credit system as *the mother of all insane forms*" (Harvey, 2014, p. 239).

The cycle of boom and bust in real estate markets can only make sense as an opportunity for long-term investment if the wealthy can buy real estate at low foreclosure prices during the economic crash that occurs after the bubble bursts and rent it out until the market catches up after a few years. As Marx describes, "As soon as the storm is over, this paper again rises to its former level, in so far as it does not represent a business failure or swindle. Its depreciation in times of crisis serves as a potent means of centralizing fortunes" (Marx & Engels, 2010, p. 468). However, these economic crises can generate deep social unrest and drastic deterioration in the conditions in which the majority of the population lives, as noted by Stiglitz in his comments on the perils of unregulated markets in Latin America:

> We recognize that the extended periods of unemployment, the persistent high levels of inequality, and the pervasive poverty and squalor in

much of Latin America has had a disastrous effect on social cohesion, and been a contributing force to the high and rising levels of violence there.

(Stiglitz, 2001, p. x)

If it was the deregulation of markets that brought us to the financial crisis, it is worth considering Polanyi's point that a self-regulating market was a novel idea up to the time of the industrial revolution, as markets and regulations had been thought to have developed simultaneously (2001, p. 72). Therefore, the assumption that a market economy that was controlled, regulated, and directed by market prices via self-regulating mechanisms

> was grounded on the expectation that human behavior was oriented to achieve maximum capital gains under any circumstances. Accordingly, markets would develop for every commodity, trade, or industry, but also for labor, land, and money whose commodity attributes would be called wages (price for the use of labor), rent (prices for the use of land), and interest (price for the use of money).
>
> (Polanyi, 2001, p. 72)

Under the above-described neoclassical approach, the markets were a vital part of the economic system in which the State should not interfere by inhibiting their formation or in the adjustment of prices under changing market conditions. As labor, land, or money are not produced for sale, describing them as commodities is rather inaccurate, and yet this assumption enabled both their purchase and sale on the market as well as the self-regulation of the system. Based on this analysis, Polanyi questioned the applicability of automatic market adjustment mechanisms to them, and called for a substantial governmental role in managing fictitious commodities through the establishment of strong regulatory institutions.

Revisiting urban land rent theories

Marx's rent theory explained how, although land is not socially produced and provides no added value, imaginary prices are established through the *capitalization of rent* "extracted" from that land. Much as the possession of capital generates interest and can be traded on the financial markets, land derives its value from its potential to generate a return, meaning that land prices are inclined to match the interest generated by the capital the landowner would receive in the future. Marx identifies three forms of rent:

1 **Primary differential rent**, which is produced out of the intrinsic features of the land and its location, in terms of its proximity to infrastructure, services, and amenities. The surplus value is the result of

institutional investment or that made by external agents, which the owner captures as capital gains through the appreciation of land values on the land market.

2   **Secondary differential rent** results from the improvements, undertaken by the owner, from which the tenants may benefit in terms of complementary services or facilities and the proximity to them, as is the case with shopping centers or office complexes.

3   **Absolute rent**, in which landowners extract the rent by monopolizing access to the land. In this way, landowners exercise greater control over the urban environment by concentrating real estate capital and negotiating with the State over the construction of infrastructure and amenities that may cause the value of their properties to appreciate. Absolute rent leads to the sequential occupation of the land according to the purchasing power of the consumer, showing a structural disadvantage in terms of access, wherein the population with fewer resources will be left without the ability to choose, and will, therefore, be forced to accept the residual land not subject to absolute rental practices (Harvey, 1973, p. 168).

In this context, economic rent serves to allocate resources efficiently, with competition mechanisms enabling access to the most profitable land uses, based on the calculation of the three types of rent described above, which concentrate wealth in certain areas of the city. The notion of land rent, as explored by Marx, has a particular rationale, identifying certain factors such as location, accessibility, or density provisions that may increase its potential for development, that may later intervene in the appreciation of the price of urban land and therefore should be considered. Other dynamics, such as technical innovations in production, gentrification mechanisms, or the construction of large infrastructure projects, also alter land prices, which may fluctuate over time at moments of an excess of said dynamics, leading to the establishment of new regulations or the bursting of real estate market bubbles. Therefore, land prices change over time, transforming the physical configuration of cities, personal interaction, and social dynamics.

How this actually occurs may also depend on questions of scale, as urban land prices respond to structural and conjunctural factors (Jaramillo González, 2009). One structural factor influencing urban land prices is their long-term tendency to increase in value, unlike most commodities.[3] Moreover, putting land at the core of the accumulation process induces a broader demand for built environments, as it brings about a corresponding growth of the urban population and an increase in wages in certain sectors of the economy, as well as changes in the particular land-use requirements of activities related to production, services, or exchange. The demand for urban space may comprise different rhythms, intensities, and scales, where, for instance, the extension of infrastructure beyond a certain distance from the existing network can increase the costs associated with accessibility, and which developers may transfer onto the local authorities or, even, residents themselves.

Urban expansion creates conditions conducive to the scarcity of land, such as the rising cost of extending infrastructure, the expansion of travel distances, and/or the reduction of available land for social use, which can each increase the cost of the reproduction of labor power. The appreciation of land value may correlate to land scarcity, and to the possibility of transforming rural fields into urban land, in the process intensifying land use through urban densification provisions (provided that rent revenue justifies the vertical expansion of construction), or public investment in infrastructure or facilities that increase the potential of the area. Other factors affecting urban land prices are related to variations in other markets, such as fluctuations in the construction industry, wherein the demand for land for development and urban expansion may vary depending on the competition mechanisms for the best locations or deals for large properties. However, the construction sector is particularly sensitive to fluctuations in other markets, variations in interest rates, or State intervention in large urban operations, such as social housing developments. In recent decades, the dynamics of the financial markets have been decisive in defining the land prices that dictate rent levels and the general rate of profit has been instrumental in setting land prices through the capitalization of rents (Jaramillo González, 2009, p. 194). An important conjunctural factor is the speculative demand for land: for instance, when the property is acquired at a certain price with the aim of selling it at a higher price and, in the process, triggers a self-fulfilling mechanism whereby prices escalate due to the belief that prices will continue to rise indefinitely, but this assumption ends when the housing bubble bursts, leading to a new process of price equilibrium.

The growth of cities brings about not only urban expansion but also changes in land use, which generates transformations in the allocation of urban space. These transformations occur as the most profitable land use invariably prevails in the market over competing land-use types, regardless of the impact this change may have on a given part of the city, as when appreciating or devaluating properties leads to the restriction of particular uses. This occurs in residential areas that begin to transition into commercial districts or when rural land begins to be absorbed by the city, transforming the differential rents previously captured for that land.

It is not necessarily the most productive land uses that colonize the areas in the city with the highest levels of rent, as with central areas of cities that are often dominated by tertiary activities such as banking, insurance, financial services, entertainment, or tourism – activities not oriented around production in the traditional sense. Therefore, the cost-effectiveness of central locations does not necessarily arise from the marginal productivity of the land but instead results from the process of capturing absolute and monopolistic rents. On the other hand, densification or changes to the urban density index produce a price increase in the long term, although this may also happen when a real estate developer passes a land price threshold that justifies the construction of higher densities on certain properties. The operation of the real estate market is affected by one particular agent, urban speculation, in which investors seek to profit from the collection of rents as well as from the increase in land prices through the trading of properties that generally appreciate in value over time, due to circumstances not produced by said investors.

Other important factors in the prediction of future land prices are information on future urban operations and construction developments, changes in urban regulations or building codes, and public investment in infrastructure or facilities. Developers or investors will seek to obtain such information through their connections with various levels of government. In this sense, major agents of capitalism may induce certain kinds of urban operations through which they can drive direct urban development for their benefit, producing both the narrative and the instruments to achieve their objectives, but which will also transform the socio-spatial structure of the city.

As land markets operate according to the principle of supply and demand, artificial scarcity can be created when landlords keep large areas of land off the market, thus creating scarcity and rising prices. In this sense, changes in the land-use structure of cities are closely related to land prices and correspond to an attempt to maximize rents, which will force current residents to migrate to a less expensive area of the city. The use of land prices to guide urban development is far from desirable, as the operation of markets does not naturally yield equity and efficiency, because they are based on abstract ideological assumptions, such as widespread market information, predictability, uniform conditions, and other conditions affected by power relations and structural conditions (Abramo, 1997). One of the undesirable outcomes of allowing market rationality to dictate urban development is the processes of hyper-densification that create congestion, pollution, segregation, sub-utilization, and the over-exploitation of natural resources. Moreover, urban sprawl, another undesirable outcome, increases the cost of extending infrastructure, thus affecting mobility and reducing the available facilities, amenities, and public spaces.

The formation of urban land prices

The establishment of land prices presents a series of paradoxes from a Marxist point of view as, while prices are characterized as the valorization by the market of the intrinsic characteristics and location of the property, what is valorized is the potential to capture rent from the use of the land. As we have pointed out, land markets are characterized by structural rigidity wherein scarcity is related to the existence of limited stock. While regulations may limit construction rights as well as the practice of specula-tive land retention by owners, the urbanization of certain areas could draw the appreciation of land prices. In this sense, land rent can be considered both as fixed capital and as the transformation of the concrete surplus value derived from the capital appreciation generated by certain land use in a given space. Such capital gains are fixed in the rental price paid by the tenant, which translates into profit for the real estate developer or bank interest for a financial institution.

For instance, we can take the case of housing. While land prices seem to be established based on consumer preferences regarding accessibility and location, significant discontinuities are observed in the price scale derived from such criteria (Topalov, 1979), creating different housing submarkets with their own particular price formation mechanisms. Such mechanisms generate heterogeneous and hierarchical urban areas, characterized by a tension between the upper and lower extremes of the real estate market and constituting one of the essential engines of urban speculation. In this regard, if the price of land determines its use, then the mechanisms that determine location operate under conditions of strong speculation that uses zoning as a rational instrument for determining the location for the organization of production and distribution. From this perspective, the spatial configuration responds to the structural conditions of the market economy, implying a series of contradictions and inequities intrinsic to the model, wherein urbanization becomes both the result and a condition of the capital circulation process (Santos, 1977, p. 10).

In short, land prices respond to transformations in the urban struc-ture due to changes in land-use provisions and construction intensity coefficients, where certain sectors of the city upgrade their land-use classification to reflect a hierarchy of value, thereby increasing land prices there. Then, if the total land rent exceeds a certain threshold, it opens up the possibility of vertical construction, laying the foundations for requesting changes in the zoning parameters and causing prices to rise rap-idly. Figure 2.1 shows the trajectory in which the price of land follows an initial trend of general growth, then increases in an accelerated manner during the change of use process, and finally returns to the previous rate of price increases.

Zoning regulations may restrict the supply of housing since they induce scarcity and raise its prices. In Fig. 2.2, A represents the equilibrium point between the number of dwellings supplied Q_E and the demanded price P_E. When zoning restrictions limit the supply Q_{Max}, the price will rise to P_{Zoning} as Q_{Max} intersects the demand curve at point B. The estimated zoning effect is estimated to be between what a buyer would pay (P_{Zoning}) and the marginal cost of supply (P_{Supply}). This price is what a developer would be eager to pay for the authorization to build one more dwelling unit in a given location.

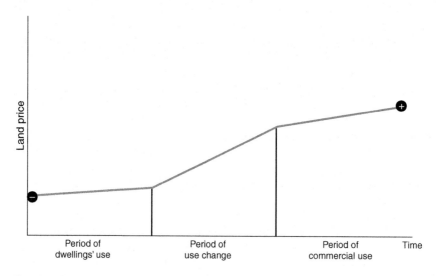

Fig. 2.1 Long-term structural variations in land prices.

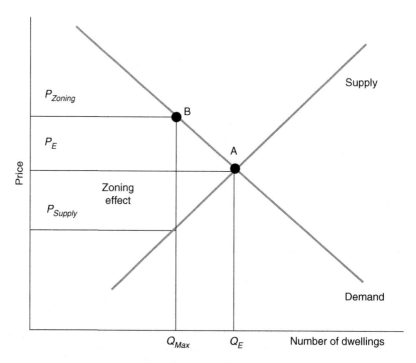

Fig. 2.2 The effect of zoning regulations on housing prices.

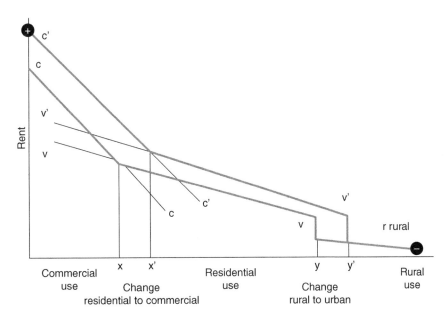

Fig. 2.3 Land-price evolution after land-use change.

To illustrate the above-discussed transformations, we will take the case of the interaction among rural, urban, residential, and commercial land use (Fig. 2.3), wherein the demand for these last two uses increases from C-C to C'-C', thereby increasing the housing rent curve from V-V to V'-V'. With this, the point of intersection of the new income curves for commerce and housing is X', where the price of commercial use displaces residential use, an increase expressed by the distance between C-C and C'-C'. In this sense, the price of housing increases according to the distance between V-V and V'-V'. Given that the price increase between the change from residential to commercial use reaches C'-C', this implies an increase greater than that observed with other land uses, thereby generating a new hierarchy within the spatial structure. This phenomenon points to one of the mechanisms of urban expansion that characterize cities in Latin America, although it is necessary to point out that, historically, this process has also been associated with the invasion of agricultural or conservation lands and, in some countries, the conversion of collective property into private.

The action of speculative agents is crucial to the formation of land prices, as they may actively intervene in the appreciation of land prices to capture the surpluses generated within the real estate market or wait for an opportunity to do so. Those individuals belonging to the intermediate circuit of the spatial economy are the end-users of properties, those who

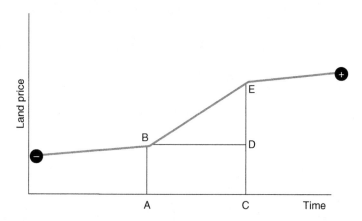

Fig. 2.4 Representation of a passive speculation operation.

possess the property for their use, and, yet, over time will capture the profit generated by increases in its value, leading to the property being characterized both as a consumer good and as an investment.

In a second variant of the above-described process, speculators buy land before prices change in order to capture the capital gains from its later sale. Figure 2.4 shows the structural price change when the buyer acquires land at point AB, which, after a certain time, can be sold at the corresponding price increase observed at point CE. In this case, the surplus value is represented as DE, while the capital turnover occurs in the period AC, capturing a profit equivalent to a productive investment without any actual manufacturing process taking place (i.e. fictitious capital). In this way, the role of the speculator/investor consists of identifying the places in and times at which these value transformations occur, taking advantage of privileged information regarding institutional interventions in the territory that may affect the appreciation of land value in the short and medium term.

There is another kind of speculation, described by Jaramillo González as *inductive* (2009, p. 129), whereby the developer buys land reserves for the construction of housing that will later require commercial spaces, triggering with it the rescaling of land prices. In some cases, part of the land can be withdrawn from the market to capture the capital gains earned by the development itself, or by making *land readjustments*, which induce changes in the price of land. In terms of vertical increases in density, a similar mechanism reduces rent to levels lower than those observed in the center of the city. In Fig. 2.5, Point A indicates the land price at which it is feasible to build vertically, while Point X indicates the location at which higher construction density can be introduced. This thus generates a sharp increase in land price, represented by segments X to X', greater

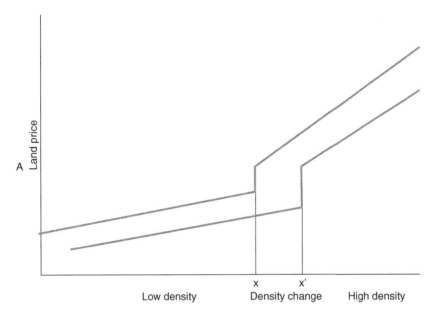

Fig. 2.5 Land price evolution after building density change.

than the increase observed in the rest of the indicators, which, at this location, reflect a secondary differential income not previously present.

Another mechanism that affects the real estate market is the limits imposed on the supply of land by keeping large tracts of land out of the market, thus creating an artificial shortage of land and raising its price. This practice is a reminder that changes in the spatial structure of cities are closely related to land prices, with State intervention being decisive in preventing developers and landowners from maximizing their rents at the expense of residents, who are forced to migrate to less expensive areas of the city, in the process feeding the mechanisms of socio-spatial segregation inherent to this logic.

It should also be noted that the State can affect the conditions in which land markets operate through the collection of land taxes, which can be designed to induce conservation, densification, urbanization, diversification, or urban regeneration practices. In this sense, urban regulations may allow certain uses or densities on the understanding that they will affect land prices, thereby presenting opportunities to capture capital gains, which can be invested for the benefit of citizens. But, as noted above, State intervention can involve large infrastructure investment in roads, electricity, water, and sewerage, etc., which has the potential to increase land prices, and thus enable the capture of capital gains, further to promoting large public-private infrastructure or social housing projects, in which the State provides a financing fund or participates in specific trusts. Despite the social scope of these measures, State intervention in this type of operation has been widely questioned, under the rationale that the control of land prices can result in inefficient land use, the over-exploitation of natural resources, or the addressing of the effects of structural economic processes rather than their causes. Therefore, the State can raise revenue via the collection of taxes and contributions, which must be fair and reasonable for public finance mechanisms to be sustainable, where the recipients of these mechanisms must be treated fairly and the taxes and contributions collected used in a redistributive fashion, using a progressive rationality or proportionality to wealth. Notable among the instruments of territorial tax collection are taxes on property value, contributions for realized profits, and the capture of the surplus value derived from public interventions.

In the last century, decentralization – whether programmed or spontaneous – has flattened the density and market value gradients identified in the central business districts when companies or housing developers start to decenter their operations and explore partially expanding markets. In this sense, companies will settle in locations where they obtain the greatest return on their investment via lower costs. However, it is not always easy to change locations, given the high level of corporate inertia observed, except in the retail sector, whose efficiency is highly dependent upon location, which is essential for successful market segmentation or niche marketing strategies.

The intended function of price mechanisms and the market as rationing devices does not exempt them from the critique that land is a commodity, which should not be subject to central or local government planning controls. Offsets for property rights expropriated by the State may be controversial regarding the value of developments that are subject to taxation (also known as improvement contributions). That critique is directed at planning authorities' ability to create scarcity mechanisms through the restriction of land use, zoning density regulations, enhancement charges, and expropriation or levy provisions. In this regard, the general principles of taxation in the urban sphere must dictate an equitable division of the profits derived from the development between the treasury, the local government, and the developer or investor, in such a way that taxes do not hinder the undertaking of urban operations that attain the public interest.

After revisiting the classic spatial economic models and the economic structure of cities, a final theoretical framework will be advanced on the three circuits of the spatial economy, started with Christaller's hexagon marketing principle that Milton Santos used to identify two circuits of the urban economy in developing countries and from which a third circuit will be presented also for this region.

The three circuits of the spatial economy

The city can be understood as a set of circuits that comprise the spatial economy and which can be seen as structured through dynamic relationships, each having its own socio-spatial-temporal structure. But, as Durkheim pointed out, stratification is an integrating structure of the social system, and such social systems, once they stabilize, engender an accepted, and subsequently institutionalized, pattern of social stratification (Parsons, 1951, p. 266). Studying these structures requires analysis of the relationships among the elements that comprise them, where, in the neoliberal city, space is a system produced by capital and monopolized by dominant groups that organically generates mechanisms of segregation through the operation of market forces. Given this, the political-legal structure expresses the struggle for resources and power through the division and segmentation of space, which present contradictions, conflict, and alienation, as Borja and Castells (1997a, p. 127) suggest.

According to Milton Santos, the spatial location of land use in developing countries results from a combination of elements that respond to a global matrix of modernizing forces that dictate the spatial differentiation and specificity of places (1977, p. 49). However, the effect of such forces often alters the stability of the spatial organization, leading to segregation, polarization, and differentiation within the urban structure. Santos introduced the idea of two overlapping urban economic circuits, the traditional and modern, which call into question Christaller's concept of threshold and Lösch's demand cone, where "contrary to developed countries, consumption possibilities are unevenly and not extensively diffused but, rather, concentrated in certain points, as a partial consequence of income disparities of the distribution system" (Santos, 1977, p. 56). In this sense, both circuits compete for the market, wherein the upper circuit prevails and increases its share and the lower circuit creates a new role in the spatial organization of the city. To a certain extent, the expansion of one circuit takes place at the expense of the other, in response to the geographical distribution of income, transport facilities, and economic activities.

Notwithstanding, Santos's two-circuit model of the spatial economy leaves out a circuit that is fundamental to understanding the mechanisms of reproduction for the spatial structure – namely the intermediate circuit of the spatial economy (1977, p. 49). This third circuit is the formation of the foundation of the spatial structure through the establishment of a formal territorial base that is neither directly linked to the financial capital invested in the real estate market of the upper circuit nor the informal urban economy that characterizes the lower circuit, therefore establishing an alternative dimension of the formal real estate market (Fig. 2.6).

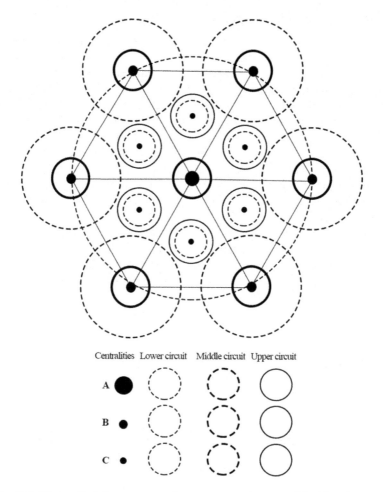

Fig. 2.6 Christaller's hexagon marketing principle as modified by the introduction of the three circuits of the urban economy in developing countries.
Source: Author after Santos, 1977.

This configuration of three spatial economic circuits reflects the socioeconomic divisions of the population within a spatial structure comprising the following: a privileged upper circuit, linked to modern technology and high income levels; an intermediate circuit, in which the daily life of the formal city is managed; and a lower circuit characterized by an informal economy that integrates the most disadvantaged population. These circuits are hierarchized through their prices and production mechanisms, forming particular real estate market segments.

The above-discussed circuits compete for both market and spatial control, in which, while solidarity is functional and antagonism is structural, they maintain a reciprocal and interdependent cause-and-effect relationship. There are specificities of place that correspond to each circuit, which are spatially different as a result of historical variations, producing conditions of imbalance, instability, readjustment, the adjustment of forces, and multi-polarization in the spatial structure of cities. Thus, a hierarchy of economic activities is generated, as consumption capacity varies according to spatial location. In this way, the three circuits intersect at points and times determined by reciprocal relationships.

The upper circuit

In the upper circuit, the inherent scarcity of land is derived from the peculiarities of location, which are not interchangeable, thus generating a diversification of prices through monopoly. In this circuit, which is based on both intrinsic differentiation and relatively uniform consumption, consumers seek to find elements of distinction (Bourdieu, 1984), where a set of generative schemes conditions the individual to perceive the world differently. Symbols of social status, such as the natural characteristics of the site, sports equipment (such as that required by golf and tennis clubs, etc.), or extensive surveillance, are factors that increase a property's potential to appreciate in value.

The price of residential housing in the upper circuit is isolated from the regulatory reference price in such a way that it no longer depends on production costs and relies instead on the monopoly price, thus creating hierarchically organized submarkets. The monopoly price results from the scarcity of locations, progressively restricting space while maintaining differentiation based on the symbols of social status; materiality is expressed through the consumption of the goods available in the upper circuit.

It should be noted that while the upper circuit presents great residential stability and little interurban mobility, a sharp decrease can sometimes be observed in the residential population of the high-income districts of the center of the city. This movement is typically the result of commercial and financial activities competing for space in conditions where the latter have superior purchasing power. In this circuit, real estate developers are frequently found on the margins of market expansion, as the high price of land makes an investment with a good rate of return unviable. This forces them to find new undervalued urban areas, which can be developed through large-scale urban projects and operations or the eviction of residents of neighborhoods subject to recycling, rehabilitation, and urban renewal projects.

Real estate capital is concentrated in this circuit: developers, builders, financiers, tenants, and real estate agents function as the main agents. A series of direct links can be found here, as international financial entities, the stock markets, and developers work on all four continents and circulate indefinitely between global and local spaces. The spatial structure of cities reflects the logic of the market and is configured to respond to its demands, serving as material support for financial globalization and the information and technological revolution, generating discontinuous and self-contained spaces segregated on a metropolitan and global scale.

The intermediate circuit

According to Santos, the upper and lower circuits are connected by the middle classes (1977, p. 59), which we propose as populating the intermediate circuit of the spatial economy. This circuit operates within the framework of the formal economy without access to financial capital on a large scale, except through the stock exchange mechanisms used by large private developers dedicated to the mass production of social housing who use mortgages as financial assets, as seen in Brazil, Chile, Mexico, and Colombia. In this regard, it is important to note that, for a new housing market to exist, a secondary mortgage market must be activated to enable a segment of the population to access second-hand housing, which, in turn, gives the seller access to the capital required to acquire a new property, thus forming a mechanism for the circulation of capital.

Analyzing the formal housing market in France, Pierre Bourdieu (2003) makes a specific criticism of the economic practices that respond to the individual's experience of the regularities of the real estate market rather than acknowledging a rational decision based on the valuation of a property. Thus, decisions related to housing access may respond to economic provisions, tastes, access to credit, subsidies, social protection networks, or the conditions of supply in the housing market, in addition to the symbolic value of the household to the family.

The intermediate circuit also comprises housing rental schemes, where rent is valued as either capital or as the capitalization of net income and is equivalent to the price of the right to use a property for a particular period of time. Within this modality, the unit of property will remain on the market as long as it is more profitable for the owner to keep the property than allow it to degrade until its demolition, as has historically happened as a reaction to rent control policies in historic preservation districts. Therefore, from the point of view of the developers, the location of the property must enable it to remain in the market, while the renovation of their property would give access to a market that yields higher rents, as happens with the gentrification of historic centers.

The lower circuit

In the case of cities in developing countries, the *lower circuit* comprises an extensive network within the spatial structure where the informal sectors are configured as a shadow economy that intersects with the upper and intermediate circuits in a formally unregulated environment. Moreover, given that real estate markets operate under regulatory parameters, and although the land and property characteristics of this circuit do not have formal property titles, there are other ways of legitimizing the ownership of the property through certificates, private contracts, assignments, and receipts, etc. In this context, even though the informal housing sub-market operates with certain limitations, properties do not generate sufficient profit to attract formal capital investment. Moreover, consumers in this sector are unlikely to have access to the credit or mortgage lending that would give them access to the financial system and, therefore, speed up the return of their capital.

Self-construction remains an important segment of the informal housing market and, although legal uncertainty regarding property rights prevails, transactions take place on a regular basis among the low-income population. In the lower circuit, territorialized economic activities may even correspond to economies of scale, such as the trade of food and raw materials, mass transportation, the informal construction industry, and services for the population with scarce resources.[4] The prevailing logic operating in this circuit dictates that out of necessity the resident population fights back through mechanisms of social protest, mobilizations, and, ultimately, political patronage. A principle of proximity related to a socioeconomic and cultural identity is also observed in this context, as is the potential for the future valorization of location.

The lower circuit, therefore, maintains a direct economic correlation with the other two circuits, either as a provider of labor for maintenance, construction, and services, or as buyers of consumer products, thereby creating an economy of reciprocity. This circuit is the point through which the disadvantaged population can access the city and it is organized into a framework, equivalent to its formal counterpart, in which participants operate mechanisms of land ownership with different degrees of legitimacy. These mechanisms have, in recent decades, considerably increased the rent levels levied on homes in these irregular settlements, giving rise to new spatial configurations and a weakening of territorial appropriation. Therefore, the lower circuit depends not on formal real estate production mechanisms or existing financial mechanisms, but rather on the relationship between local supply and demand, which is characterized by a structural deficit of housing stock as well as underestimation of its value, given the more marginal conditions found there compared to the rest of the city. This prompts questions as to how urban space is organized, based on the differences in locations, and how land prices are formed. As discussed below, the answers to these questions relate to the criteria for spatial distribution, the class conditions, and the practices employed in the urban space by social agents.

The spaces surrounding those dedicated to financial capital in the upper circuit result from the material production processes taking place there, where a series of privileges in terms of infrastructure and public facilities and services operates. These privileges are linked to the functions of finan-cial capital, producing spatial segregation and leading to the qualitative transformation of public, educational, and cultural facilities. Therefore, regularities in terms of the social differentiation of space cannot be attrib-uted to intrinsic spatial properties but instead to internal structures of the capitalist production system. These structures represent deep/struc-tural tendencies for the transformation or appropriation of those spaces representing the greatest possibilities for maximizing the profitability of activities.

It is important to realize that the formation of spatial structures in cities depends, to a certain extent, on the action of the State and, increasingly, on real estate agents who have a definitive impact on both the location of different social classes and land-use changes. In particular, the crisis of real estate overproduction has led to new urban operations in areas that were expected to experience a transformation in the dynamics of land prices, which have led to the revaluation of the area and enabled the capture of the profits derived from the intervention. These operations include activ-ities such as urban renewal programs in historic districts, the construction of residential developments, or the undertaking of public infrastructure, services, and equipment projects.

To justify this type of intervention, public policy uses economic theory to legitimize the actions undertaken by agents and institutions seeking to promote a particular development model. In this sense, public policy follows a system of beliefs and values associated with an "economic common sense linked, as such, to the social and cognitive structures of a particular order" (Bourdieu, 2003, p. 24). Hence, the logic of the real estate market, and, therefore, of private developers, does not follow a ratio-nale for reducing social cost, wherein those areas best served in terms of infrastructure would be sought. Instead, private developers are governed by the logic of maximizing the rate of profit derived from the appreciation of land values. As a result, lower-income populations, generally living in informal settlements, social housing developments, or deteriorated his-torical centers, are prone to generating a sense of resistance toward the transformation of their living environment.

According to Harvey (1973), business cycles generate over-accumulation mechanisms that produce a flow of investment into urban space and create the basis for future accumulation. As Lefebvre comments, "while the degree of global capital gains formed and realized by the industry decreases, the degree of capital gains formed and realized in spec-ulation and through real estate construction grows. The second circuit supplants the main one. From accidental it becomes essential" (Lefebvre, 1970, p. 165). Just as Lefebvre points out with regard to the transference of circuits in the secondary circuit of capital, the moment finance cap-ital begins to dominate the real estate market it will tend to suppress the imbalance between supply and demand to establish a monopoly price.[5]

Thus, when entering the real estate market, financial capital will tend to control its orientation, accelerate its concentration, and centralize capital gains.

Nonetheless, when financial surpluses are only invested in certain areas of the city as a spatial fix, this can lead to the over-accumulation of capital. Because the general rate of profit is fundamental to the definition of land prices through the capitalization of rents (Jaramillo González, 2009, p. 194), an important conjunctural factor is a speculative demand for land, which seeks to benefit from the appreciation in value in the short and medium term. The extensive demand for land can produce a shortage that can lead to the development of a real estate bubble, wherein expansive demand is generated for a *commodity*, being the case of land, whose price does not respond to the real demand but is linked, instead, to the fictitious price increase created by speculators.

Finance capital ends up directly controlling the real estate development sectors that deal with large corporate centers, property, and even already privatized land uses and public services. In this way, the frontier of urban development is not established by any official norm and is based, instead, on the general conditions relating to the valorization of space, in the absence of which private capital will not undertake any project.

The dynamics of the real estate market are directly related to urban expansion and changes in land use and intensity. Unless the State intervenes in these processes, the most profitable use will prevail, displacing essential inclusion and the equity factor to achieve territorial balance. Such is the case when rural land is transformed into urban land, when residential use is transitioned to commercial use, or when social housing changes category via gentrification mechanisms, in the process transforming the differential rents that it previously generated. It should be noted that the most productive uses do not always secure locations in the most profitable areas in the city, if finance, banking, entertainment, or tourism services have become the predominant land uses. Therefore, contrary to neoclassical theories, the profitability of central locations does not necessarily originate from the marginal productivity of the land but instead results from the process of capturing absolute and monopolistic rents. The processes involved in densification coefficients or changes in land use produce an increase in prices, provided the developer can identify a threshold in land prices, above which the construction of higher densities in a given location is justified.

Social and institutional frameworks

In the last quarter of a century, real estate markets in Latin America have been transformed by the entry of developers associated with financial capital, leading to an exponential rise in urban land prices. This phenomenon is explained by the introduction of speculative capital through financial instruments that generate mechanisms for expansion and building intensification, independent of market demand, and leading to increased rents, residential segregation, hyper-densification, as well as the saturation of infrastructure and local services.

The emergence of financial capital in the real estate market is only made possible via State engagement, establishing the operational bases for large-scale urban interventions through regeneration projects in historic centers, large corporate, commercial, and residential complexes, and housing developments. The government assumes the role of generating a favorable business climate for these interventions. As Capel suggests,

> The current situation shows ways in which [financial] capital is projected into the city through large projects or investments that profit from all the advantages that the public apparatus offers, being in charge of land-use regulations, a feature that explains their great flexibility to accommodate investments from residential projects.
>
> (Capel, 2002, p. 145)

In this context, the so-called *capture* of the State takes place (Kaufmann & Hellman, 2001), wherein public agencies are the dominant economic power group that, using their political networks, can impose the interests of financial capital on the territory through plans, projects, and even public-private partnerships that guarantee institutional support over time.

The State has always played a fundamental role in the development of the real estate market, be it by action or omission, using different means and mechanisms. Traditionally, the government has been a promoter of urban space, building public infrastructures for administrative, cultural, educational, and health uses, and partnering with the private sector in large-scale housing developments, infrastructure, and services projects. As an important agent of territorial transformation with a social commitment to the population, the State, we will later propose, is entitled to capture the capital gains obtained from price increases generated by these types of operations, through fees, taxes, and even negotiations with the owners and developers who benefit from State intervention.

The need for an institutional intervention in the real estate market is based on the territorial imbalances derived from the unregulated operation of market forces, which trigger mechanisms of socio-spatial segregation, the underutilization of land, the saturation of infrastructure, hyper-densification, the obsolescence of real estate stock, and speculative practices. In response to these mechanisms, the State may adopt policies to guarantee spatial balance, equity, functionality, and the redistribution of the assets produced by the community.

In service to the foregoing, the State and its institutions have the legitimate right to use physical force, either through the police or military forces, which is derived from a process of accumulation and concentration of different types of capital (cultural, social, symbolic, and economic), and this may have a determining impact on the economic functioning of the urban space. Moreover, the State also exerts a decisive influence on the balance of political and social forces, guiding their intervention to achieve certain objectives of its government. In addition, its influence extends to consumption and the population's living standards through the structural impacts of its laws and regulations, public budgets, and interventions in the construction of infrastructure and services.

While land markets may operate independently, the State can affect the conditions in which they operate through land taxes and contributions, where conservation, densification, urbanization, diversification, or regeneration practices can be induced through urban provisions that allow or restrict certain land uses and densities to model land prices. Traditionally, State intervention has provided large-scale infrastructure, such as roads, electricity, water, and sewerage, considerably increasing land prices, and, in recent decades, it has adopted an entrepreneurial role as a developer, especially through its participation as a major landowner in large-scale social housing projects or with the use of public-private partnerships in large-scale urban operations. These kinds of interventions have been widely debated, with the argument made that control (or the abolition thereof under socialist regimes) over land prices may result in natural resources being misused or wasted, or even that the consequences of structural economic processes will be addressed rather than their causes, as noted above.

Urban land has been subject to the rationale of the market's optimal allocation of scarce resources because its stock is fixed, but the legal claim to the land is a condition for the valorization of capital. In this sense, the urban structure materializes the economic divisions of space that result from the capitalist mode of production in the context of historically determined social configurations (Lipietz, 1985). These mechanisms contribute to the reproduction of socioeconomic formations while exacerbating their contradictions, in which spatial divisions are engineered through political and urban planning practices that have been instrumental in establishing a particular spatial order that determines the economic and social uses of land, later legitimized by zoning provisions.

As a major mechanism for capturing resources, the government has used taxes and contributions, which should be fair and reasonable for taxpayers to maintain sustainable public finance mechanisms. For this to happen taxpayers must be treated equally and the taxes and contributions they pay should be used to redistributive effect. Existing taxes levied on the appreciation of land value include taxes on site value, realized gains, and development value. It should be noted that compensation for property rights expropriated by the State and the taxation of development value (also called betterment contributions) may be subject to controversy. As part of the general principles of taxation applied in the urban realm, there should be an equitable division of the profit generated by development among the treasury, local government, and the developer/investor, ensuring that taxation does not hinder the development of urban operations if they are proven to benefit the public good.

Real estate markets in Latin America

Land markets have unique conditions within the spatial economy, as they are not organized around a central exchange and total supply is fixed and cannot be reproduced, despite comprising both subjacent and superficial support. This is why land remains the least flexible variable in the system via which capital and production intersect. Moreover, land may be subject to restrictions on or interventions in the spatial structure, such as public infrastructure projects, urban renovation, or historic preservation programs that alter its market value.

Cities contain a complex mix of land uses that compete to secure the best locations and sites according to their accessibility and complementarity within the urban structure. However, to benefit from these characteristics, real estate agents depend on their economic capabilities and the convenience of trade-offs in terms of location, accessibility, and built space. The general conditions of this market are subject to the continuous adjustment of forces, of which urban expansion is a major driver for the transformation of spatial patterns of land use, intensities, and values.

As a general principle, most urban land uses will try to benefit from maximum accessibility that may improve the productivity or profitability qualities of space of certain areas of the city, especially those that would increase in value when used more profitably. Corporation headquarters, financial services, and commercial activities are located in high-density areas of the central business districts of major cities in developed countries. Meanwhile, the central districts of Latin American cities are characterized as historic areas subject to conservation regulations and with prevalent retail commerce, along with tourism services, warehouses, workshops, entertainment venues, museums, office facilities, educational facilities, and middle-income housing. However, in recent decades, these areas have seen the functional relocation of certain activities, whereby commercial, industrial, and residential land uses have been dispersed to other subcenters of the city.

A variable asset for middle-income and low-income residents is access to the jobs, products, services, and economic opportunities that are concentrated in large urban agglomerations, thus requiring a public transport system, amenities, and services, given that complementary land uses are found close to each other. According to Balchin et al., patterns of urban land use depend on accessibility, on which land-use value and intensity depend, meaning that potential investors pay a higher market price to develop a property to its full potential, as urban property has two basic values: "the capital value of buildings and sites in their existing use and the capital value of cleared sites in their best alternative use" (Balchin et al., 1995, p. 24). Therefore, spatial economics becomes a more complex, interdependent, and multidimensional framework, in which accessibility conveys greater marginal advantages in terms of distance, time, and convenience that may yield higher returns.

Various theories posit that centripetal forces of attraction and centrifugal forces of dispersion affect the spatial structure of the city, differentiating space and creating particular urban patterns. Furthermore, areas of transition may also emerge, in which land uses and activities are mixed and interdependent. Therefore, the spatial economy produces rational patterns of land use in every city regardless of original location or size, with such configurations relating, at least partially, to the maximization of profits, utilities, or capital gains via the forces of supply and demand. Hence, the greater the benefit or utility that a site may offer, the higher the rent or price the consumer will be willing to pay.

A distinctive feature of land markets in Latin American cities is the coexistence of private property schemes and social property frameworks, which bring about further complexities when the market rationale is confronted by a multidimensional configuration operating simultaneously and hindering the creation of market value in cities. This provides a different perspective on the assumption that the total supply of land is fixed while the supply of land for different uses can be increased or decreased when land use changes from agricultural to urban, private to public, and residential to office or retail. In those Latin American cities that have suffered from systemic economic crises and drastically increased inflation, property has been regarded as a *hedge* against such inflation and the atomization of savings/earnings, becoming an attractive prospect for developers and speculators, whose demand for land is affected by credit availability, population growth, economic affluence, and inflation.

One of the major factors contributing to the imbalances observed in the region has been the availability of unregulated financial instruments, which generated the subprime mortgage crisis of 2008 in the United States and later had a major impact on Latin American cities in subsequent years. This crisis saw real estate markets allocate scarce resources among competing agents, showing that they were the type of market most susceptible to change in terms of its underlying conditions, making it highly inefficient. The unique nature of the market discussed above, wherein it provides imperfect information/knowledge to both buyers and owners, makes the latter reluctant to sell even when able to secure their desired profit. Other issues specific to this market include its lengthy and costly administrative procedures, expensive legal services, and long construction processes, which are dependent on planning authorities and regulations that might prevent the most profitable land use. Also, land-use changes take place over long periods, so the market seldom reaches a state of equilibrium in which all resources are optimally used.

The impact of finance capital on the urban structure of cities in Latin America is difficult to quantify. However, the scale of such real estate investments is changing the urban profile of the continent at tremendous speed, entailing high levels of resource consumption (water, space, communications, etc.), a saturation of infrastructure and services, increased costs, and the underutilization of land. On the other hand, the upper circuit of the spatial economy generates high concentrations of wealth in specific locations of the city and directly recreates a pauperized population that is distributed across the rest of the city and its dispersed periphery, thus accentuating spatial inequalities.

To understand the importance of finance capital in the region, one needs to look to the early 1990s, when Latin American countries were made to embrace structural reforms to increase the role of market forces in exchange for immediate financial aid required by the so-called Washington Consensus. However, these reforms were intended to activate their economies despite their social impact on the population since they accentuated inequality, stagnation, and high volatility of currencies in the region. Even when Latin American markets had managed to enter the global capital markets, the uneven socioeconomic results were blamed for excess optimism, or that pro-market policies were not implemented to their full extent since local capital markets had not evolved correspondingly.

Along with the structural reforms, privatization of state-owned enterprises started in the late 1980s, conducted through public offerings on local stock exchanges, which had a direct impact on capitalization and trading in the domestic stock markets. Also, it was expected that capital markets would provide cheap financing, mobilize savings, pensions, and funds efficiently to their most productive use, opening investments to more attractive opportunities and modeling the macroeconomic environment by adapting the legal and regulatory frameworks to financial markets. In the next generation of reforms, the capital markets were consolidated, broadening pools of investors and creating new financial instruments, integrating institutional insurance, pensions, and mutual funds as capital that would provide long-term finance for private investors.

Since the expected results were meager compared to the initial expectations, international banking institutions such as the World Bank and the International Bank for Reconstruction and Development contended that financial markets "have long gestation periods before producing visible dividends. It thus recommends letting market discipline work, while forging ahead patiently with further reform implementation efforts" (De la Torre & Schmukler, 2007, p. 20). Therefore, failure to comply with expectations was not considered to be the result of bad economic policies, and yet, after four decades, the course of action was the continuation, as well as the deepening, of those same policies.

The increase in securities markets in developed countries is credited with having furthered not only financial liberalization and deregulation, but also financial and technological innovations to create an ongoing investor base. The participation of Latin American companies in international equity markets started as cross listing in the US market through American Depositary Receipts. In the case of structural finance, markets experienced a strong growth in Latin America, where the volume of transactions increased from USD 65 million in 2000 to USD 4.9 billion in 2005, and USD 14 billion in 2019, concentrating securities in mortgages, as well as in the construction sector, toll roads, and securitized account receivables.

As has been thoroughly demonstrated by Piketty (2014) and Stiglitz (2012), inequality rose sharply during the last two decades, as a sector of the population dramatically increased its wealth in part through purchases of securities that marked changes in the tolerance for risk-taking investment. Also, deregulation of financial systems allowed international banking institutions such as BBV, Santander, HSBC, Scotia Bank, etc. to establish local branches and subsidiaries in Latin American countries, which enables the connection to global markets.

Furthermore, securities markets reduced the cost of capital by inducing competition among the commercial banking sector and leading to a more efficient valuation of assets, which lowered transaction costs and increased economic savings investments. The comprehensive pension system's reforms led to its privatization, imposing mandatory, privately administered, and contribution-defined pension funds in Latin America, enabling the development of local markets by making available long-term capital to the private sector. However, capital markets for private securities have been highly concentrated in Latin America, since just a few companies participate in domestic bonds and stock markets, where the top ten companies concentrate 50% of the value traded. Although these new financial instruments and hedging practices dealing with derivatives were effective mechanisms to increase profits, they also represented high-risk practices, which expanded exponentially during the financial crisis.

Notes

1 Commodities are goods that have value, utility, and require a low level of processing; are commercialized without qualitative differentiation; and are basic components (raw materials) of more complex products. They are commercialized in spot or futures markets. The concept of commodity has been extended to any consumer good, enabling the inclusion of financial assets, such as foreign currency, stock indexes, interest rates, or reference rates.
2 Durand (2017) notes that economist Hyman Minsky made a similar point.

3 Most commodities tend to depreciate over time due to mass production, inno-vation, or competition mechanisms, with a decline in the rate of profit taking place over the long term.

4 In Latin America, the informal sector employs the equivalent of 53.1% of the population.

5 Also, together with the automation and technification processes involved in economic production, the foundations of spatial fragmentation are devel-oped in the following locations: corporate management activities take place in the privileged sectors of international metropolises; operational management functions are carried out in lower-ranking cities; and the workforce is located in undervalued areas of the city or even in developing countries.

3 Financial instruments as tools

Financial instruments and the location of capital

Spatial structures arise from the concentration of social forces generating specific internal dynamics that respond to administrative and bureaucratic practices, and in which values are culturally and historically grounded and correspond to particular spatial patterns. Therefore, it can be argued that space is socially produced, where daily practices and interactions determine specific space functions, and that cultural significance configures its particular structures (Lefebvre, 1974, pp. 39–40). These configurations respond to ongoing dynamics of financial capital, which is shaping urban space through corporative business districts, large shopping centers, gated communities, large social housing developments, major infrastructure projects, high-rise urban corridors, privately developed low-income housing, and tourist-oriented condominiums and hotels.

It is important to point out that this taxonomy is not intrinsic to its spatial features, but rather, it assembles certain functions and activities that allow the interaction, circulation, and expansion of capital. These new spatial configurations may create new centralities through economic instruments that advance finance, in which the State participates as an active agent in structuring the framework for enhancing the market's ability to generate profit. In this sense, Foucault suggested that cities had been long acknowledged as serving as testing grounds for governmental rationality before it was extended to the entire territory (2012, p. 142). Similarly, financial capital has found a *spatial fix* as an adequate mechanism to absorb surplus capital into the city, and while boosting real estate markets, capital circulates with little constraints, creating a global circuit in which assembling, producing, circulating, and consuming take place, as well as the establishment of structural dynamics according to a general financial rationale (Braga, 1997, p. 196).

The spatial structure of global cities is a physical manifestation of the financial and productive configuration of capital and their dispersed concentrations of hubs and networks within the new economic framework. Nevertheless, this world city network (Taylor, 2004) should not

DOI: 10.1201/9781003119340-4

be considered as a strategy for achieving regional growth or comprehensive economic development since its configuration is produced by financial mechanisms that recreate asymmetries and generate spatial patterns, which are not exempt from social conflict and power struggles. The new geographical organization of capitalism is advanced as a narrative in which cities seek to blend into the processes of capital appreciation, as the *anchors* for financial flows that articulate the global space profit from financial instruments that are fine-tuned for capital use (Brenner, 2009; Jessop, 2004). Nowadays, cities aiming to consolidate their position within global markets require the State to deliver operational facilities to the private sector, including access to infrastructure and quality services, a growing internal market, and tax incentives, as well as interconnecting nodal spaces located in the privileged parts of the cities. In this respect, Borja and Castells contend that world cities are ranked according to economic hierarchies serving specific functions:

> The global city is a network of urban nodes, at differing levels and with differing functions, that spreads over the entire planet and functions as the nerve center of the new economy, in an interactive system of variable geometry to which companies and cities must constantly and flexibly adapt.
>
> (Borja and Castells, 1997b, p. 61)

Even when this framework marks a global trend of accelerating financial flows throughout cities, the perspective of horizontal networks is questioned by the existence of great disparities and economic polarization, as can be seen in recent decades, which have been characterized by increasing inequalities (Stiglitz, 2012; Piketty 2014). But while financial capital has played a central role, it has generally been supported by the State, strengthening the mechanisms of exclusion, segregation, and concentration of wealth and privileges configured as *devices* of globalization (Valenzuela Aguilera, 2013).

Many of those who benefit from the over-accumulation of capital that results from productive forces seek more profitable investment alternatives through speculative mechanisms such as real estate markets. These investments, even if implying greater risks, may enhance capital gains through incorporating large-scale real estate ventures and shaping the way in which the entities in charge of providing social housing and regulating urban land work. Hence, institutional investors such as housing mortgage funds, pension funds, hedge funds, mutual funds, insurance companies, and banking institutions, among others, have channeled investments into the stock and financial markets, where both real estate and mortgages are securitized.

This is how these investors transformed real estate products into financial assets providing the required liquidity to trade in the stock markets.

As a result, financial capital began to predominate in the production of urban space through major urban operations, infrastructure, and facilities. Moreover, the process of financialization of real estate markets generated, along with the new demand for housing and residential areas in the 1950s and 1960s, led the search for new real estate products to invest in, such as large corporate business centers, shopping centers, large urban projects, international airports, sports stadiums, cultural entertainment, medical and educational complexes, as well as infrastructure projects at various scales.

It is important to highlight that these real estate products have a major symbolic significance since they transform the urban profile of cities out of an aura of modernity that allows them to compete with other metropolises of the world. In this context, historical differences, segregation, and inequalities prevail within the spatial structure, providing the opportunity to recapture the price structure that had been registered in certain areas of the city and extend the prestige of privileged neighborhoods to adjacent areas through a process of substitution of land uses, confirming the capacity of markets to reallocate the different social strata in the city even more than the State itself. Since both investment decisions and urban operations seek maximum financial performance, it is important to understand the territorial mechanisms through which financial investments operate, since they may not respond to comprehensive planning schemes that serve the needs of its citizens, favoring instead the developers' interests.

In this regard, it is also crucial to understand the process by which financial capital is increasingly becoming a dominant player, shaping market practices, narratives, agents, and actors. In order to understand the nature of this new configuration we have to acknowledge the role of the State in the creation and promotion of markets, as noted above, but also becoming a major player in establishing mortgage securitization schemes and enabling private equity funds as well as other financial entities to transform foreclosed single-family homes into real estate assets after the subprime crisis (Immergluck & Law, 2014; Fields, 2018).

Developers and construction companies no longer just use finance to fund their work, but they are themselves serving the financial network, as they are increasingly owned by multinational real estate funds (Aalbers, 2019b). These consortiums specialize in corporate business and commercial real estate,[1] but are also acquiring rental housing units. These make up large portfolios that are leveraged on investments by institutional funds (pension funds, insurance companies, etc.); their operations demonstrate how the built environment has become instrumental for creating and storing surplus value. The financial spin has been possible because the State has taken a leading role in developing real estate, where urban policy has become instrumental for establishing a rationale for financial markets, therefore shaping the urban structure of cities through fostering securitization and real estate investment trusts, so that even if decision-making is being transferred to private entities, political State power is also extended through the use of financial mechanisms.

But before turning to a discussion of these financial vehicles and their role in real estate, it is worth examining how we came to this point, a process that can be referred to as financialization. Financial capital has been transforming the economy as much as the State is, shaping with it the spatial structure and undertaking a central role in place-making when locating *spatial-temporal fixes* in cities (Harvey, 1982). In the last four decades, financialization has been framed as a systemic transformation of capitalism, creating the conditions to restructure the economy while maintaining profitability, but at the expense of high social costs when crises arise. This process has been endorsed by governments eager to deregulate and strengthen the finance-based market system fostering cross-border flows of capital, hedging risks, and setting rules among financial institutions, funds, and deposits within the financial sector.

These mechanisms have been extended to individuals and households that are increasingly dependent on the financial system to access basic needs such as housing, health, and education. Finance capital privileges circulation over production and serves as a mechanism to absorb the surplus that otherwise would lead to the stagnation of the productive sector. However, this trend has had a historical background, where the expansion of trade in the capitalist modern era experienced the rise of finance capital when production and trade declined (Braudel, 1982, p. 246). Nevertheless, financialization represents a later structural transformation of advanced capital economies, having a predatory dimension whenever financial institutions aim at systematic extraction of financial profits from individuals, firms, and institutions through market mechanisms, in which "financial expropriation is a characteristic feature of financialization and represents the restrengthening of the predatory outlook of finance toward the economy and society" (Lapavitsas, 2013, p. 800).

Looking back, we can recall the Washington Consensus in the late 1980s in which harsh economic structural reforms were recommended by prominent financial institutions such as the International Monetary Fund, the World Bank, and the US Treasury to increase the role of market forces in exchange for immediate financial help, enforcing free-market policies in Latin American countries followed by their integration as subordinates within the global financial markets due to the hierarchical organization of global capital. From a Marxist perspective, capital accumulation has been restructuring across economies of scale, expanding production through international finance, and using the State as an agent for establishing economic arrangements that have an impact on the urban structure.

Financialization has been associated with enabling the restructuring of capital through neoliberal policies, containing wages, undermining broader social conditions, and increasing economic inequality (Fine,

2014). International agencies and banks have heralded financial market development but they also recommended that careful attention be paid to the associated risks. In sum, those risks are part of a financialization rationale, which has been governing economic and social restructuring in which large sectors of the population have been affected by cyclic crises (Minsky, 1982), and in the case of the subprime mortgage crisis, experiencing considerable social downgrading.

The underlying rationale behind structured finance is to transform fixed assets into liquid – and allegedly less risky – ones, creating longer-term funding for domestic and cross-border markets that advances domestic bond markets and enhances local residential mortgages and consumer loans. The process of securitization consists of institutions assembling assets into pools of larger marketable portfolios to be sold to investors as securities through an off-balance-sheet process known as a *special-purpose vehicle* (SPV) or through a trust that funds the acquisition of assets by issuing securities to investors while holding those resources in trust. At this point, the principal and the interest of the underlying assets are managed by a servicer and channeled through the SPV, which may also hedge risks against default (also known as enhancement).

In general terms, a derivative security is a contract between two parties whose payoff depends on an underlying asset, while the kind of agreement that obligates the participants to transact at an upcoming date are known as *future contracts*. These contracts are standardized forward agreements used to trade on exchanges, for which they are highly regulated by the Commodities Futures Trading Commission (CFTC). In the case of a *forward contract*, one counterpart is obliged to purchase an asset at a future point in time while the other is required to sell it, being stocks, bonds, currencies, or commodities, which are traded at a convened price. Forward contracts are traded over the counter, taking place through a network of dealers at financial institutions, while *options* are agreements between two counterparties providing one of the parties the right – but not the obligation – to transact an asset in the future.[2]

Mortgage-backed securities (MBSs)

Perhaps the best known of these instruments is the security. The financialization of mortgages started in the late sixties in the United States when *government-sponsored entities* (GSEs) such as the Federal National Mortgage Association (FNMA, or Fannie Mae[3]) became the major buyers of mortgages in the secondary market,[4] while the Government National Mortgage Association (GNMA, or Ginnie Mae) was established to manage and sell mortgages to private investors. Later, the Federal Home Loan Mortgage Corporation (FHLMC, or Freddie Mac) was created to pool loans into packages of securities, enabling the more efficient circulation of capital as well as improving the conditions of access to financial markets for a wider public through increasing fiscal exemptions, such as allowing mortgage interest to be deductible from taxable income. The GSE operations were backed up by the US government, which allowed the issuance of residential mortgage-backed securities (RMBSs) at preferential rates.

RMBSs were conceived as pools of securities to be traded in the secondary mortgage markets in the United States, where financial institutions (*originators*) sold these packages to GSEs in exchange for cash or securities. Also, originators could sell them to private investors, who financed their purchases through credit vehicles known as *real estate mortgage investment conduits* (REMICs) issuing different term positions for those securities (short/medium/long) at various rates of interest, depending on the maturity and demand for the different pools. The RMBSs, in exchange for the pools of mortgages, could be retained, sold, or refinanced through the REMICs, taking advantage of the existence of an active *forward* market for mortgage loans and, normally, conduits and originators would hedge their risk exposures in the RMBS and treasury markets.

Originators sold the pools to conduits that warehoused the loans until the portfolio was large enough to be securitized, while hedging their risk exposure in the secondary market, where mortgages were refinanced and circulated in the capital markets as RMBSs. GNMA endorsed these mortgages as corporate instruments and warranted the full and timely payment of the principal and its interests by several US government agencies. However, it is important to stress that even when the equity of Fannie Mae and Freddie Mac was owned by private investors, their congressional charter mission was to lower the cost of mortgages to low/moderate/middle-income Americans through the secondary mortgage market. To this end, both FNMA and FHLMC bought qualifying fixed and variable rate mortgages, as well as extended a net of service contracts from originators, who were responsible for collecting payments on insolvent accounts, undertaking proceedings to foreclose properties when necessary, liquidating the same to recover their maximum value, and retaining a percentage of the outstanding balance as a fee for their services.

During the eighties, the rise of interest rates forced Fannie Mae to explore variable-rate mortgages as well as a set of financial engineering mechanisms that would permit it to access other forms of capital and extend credit to lower-income groups whose solvency was compromised. Later, in the first decade of the millennium, Freddie Mac and Fannie Mae controlled nearly 60% of the mortgage market in the United States, while home values steadily increased over the years, even as interest in the secondary market was significantly higher than the primary market. At the time, the US Federal Reserve started raising short-term interest rates from 1% to 5.5%, increasing credit charges up to 50%, which, along with the commissions for anticipated amortization, rendered it impossible for borrowers to refinance their mortgages. This situation set off the possibility of general insolvency as well as the crash of the market's bubble, which led to the subprime crisis (2007–2008) that had a long-term impact on the economy, and extended to Latin American countries.

In Latin America, specifically, an important agent for the development of real estate markets were the *mortgage-backed securities* (MBSs), as they fostered the expansion of financial capital in the region, in particular Brazil, Chile, and Mexico. In the latter, a strong political drive to overcome an estimated housing shortage of 9 million units drove the expansion of the largest MBS market in the region. As in the United States, a federal mortgage institution (Sociedad Hipotecaria Federal – SHF) was established to insure mortgages through financial intermediaries, a mechanism that will be developed later in this chapter.

In order to get hold of the legal frameworks of securitization mechanisms in Latin American countries, it was expected that regulations would guarantee the transfer of underlying assets to be securitized. In this respect, Brazil and Mexico accounted for three-quarters of all domestic securities in the region, which indicates the degree of development of their financial systems, as well as the limitations of the legal frameworks for the operation of these markets in the rest of Latin America. However, it has also been suggested that structured finance complexity has hindered investors' capacity to assess the credit risks associated with this kind of financial products (such as MBSs or CDSs), as well as the different tranches of the final product,[5] which, combined with the credit assessments of the rating agencies, proved to be a liability during the subprime crisis, diminishing creditors' ability to forecast defaults on a security's underlying obligations.

According to an International Monetary Fund report, favorable financing conditions and terms of trade fueled the domestic demand for credit in Latin America, where its expansion found a welcoming venue in the mortgage market, which was addressing the housing deficit in the region (Cubeddu et al., 2012, p. 3). Although this has been the traditional narrative of financial capital development, it is a major concern that in the Mexico Population and Housing Census 2020, more than six million social housing units were reported as vacant, and yet that did not keep banks from issuing half a million new mortgages annually during the last decade.

The Latin American region has historically protected housing rights of the population, but, in the late 1990s and early 2000s, regulatory provisions and legal amendments were put in place in Mexico, Chile, and Brazil to accelerate foreclosure procedures as well as secure the effective transfer of debt obligations. The same IMF report praises the more financially integrated economies in the region, where Brazil provided subsidized mortgages through State-owned banks, while Mexico guaranteed mortgages through a government agency (SHF) created to support the MBS market (Cubeddu et al., 2012, p. 5). When considering the spatialization of finance capital in the last decades, it is important to point out the correlation between the expansion of the construction sector and the increased availability of mortgage credit, rendering it a spatial phenomenon, turning the asset securitization market into the largest sector in the US markets, with an outstanding balance exceeding one trillion dollars (Schopflocher & Manzi, 2020).

As described earlier in the chapter, the securitization of assets involves the process of reallocating mortgage-related liabilities among different groups of investors that carry credit, prepayment, and interest-rate risks. For instance, a banking institution could take on those risks, but once the assets are securitized, those liabilities are transferred to investors according to their underlying assets, through financial instruments such as mortgage-backed securities (MBSs) and asset-backed securities (ABSs) which can be sold to institutional funds such as private pensions, retirement funds, money market funds, international and domestic commercial banks, real estate investment trusts (REITs), private investment partnerships, insurance companies, corporate treasures, individual investors, and fixed-income mutual funds. These securities allow financial institutions to trade their whole portfolios to MBSs in order to reduce their exposure to risk as well as diminishing capital investments required to hold complete mortgages loans, adding with it more liquidity to their balance sheet and enabling them to increase their fee income by writing more mortgages. In this way, banks were able to secure segments of the portfolios in the secondary market acting as sellers or buyers, transforming them into liquid securities that complied with their asset location priorities, as well as meeting their risk-management standards.

To this end, whenever the mortgages did not meet the standards set by the *government-sponsored entities* (GSEs), the private-label market provided an alternative source of funds, in which MBSs offered a wide range of possible risk profiles depending on maturity, prepayment, duration, coupon, and liquidity spectrums (divided into floating, fixed, inverse-floating, zero, and inflation-indexed coupons). In such cases, MBSs yielded less profit than the loans underlying the securities because of the hedging cost and credit-enhancement mechanisms that can be used to reduce credit risk to investors. However, when institutional loans were backed up by federal sponsoring agencies, those securities were still subject to adjustable interest rates, which could be hedged with risk management features providing monthly cash flows for reinvestment at current market rates.

Sales of MBSs grew exponentially in the early 2000s since they were endorsed by federal agencies and at the time of purchase they were considered the safest type of investment. They were rated as such by a least one nationally recognized rating agency (before the subprime crisis three-quarters of MBSs and ABSs were rated as AAA). However, in the late 1980s, new forms of MBS emerged that bundled *subprime* loans, loans that did not comply with the institutional standards required for Fannie

Mae or Freddie Mac guarantees. Subprime MBSs were originated with different underwriting criteria than regular mortgages, securitized with tailored cash-flow structures, and rated with specific credit enhancement requirements, becoming one of the fastest-growing sectors in the securitization market. Many had interest rates that were fixed at a low rate for the initial two or three years and became adjustable or simply much higher for the remaining term; this developed into a major problem when interest soared during the subprime crisis because borrowers were unable refinance their loans before the end of the initial term because the credit markets shut down. The loans took a number of forms: many were *home equity loans*, first- and second-lien mortgages for homeowners to borrow against the accumulated equity of their houses; others were mortgages for borrowers who may have had a credit history of delinquency and default, and others loans for people who wanted to purchase a larger home than they might otherwise because the credit appeared to be so inexpensive or could always be refinanced before the rate went up, where two-thirds of these mortgages were for refinance and one-third for home purchases (Tavakoli, 2008, p. 165).

Residential mortgage-backed securities have been a major funding source for the residential mortgage market, and since housing is a basic human need, securitization had already an inelastic demand to meet.[6] Each country has developed different forms of housing finance entities such as commercial banks, building societies, thrifts, saving banks, and cooperative credit societies. At some point, countries experimenting with mortgage securitization and structured finance realized that mortgage lending and servicing costs had come down while becoming an efficient mechanism to turn to (Schwartz, 2009). In most cases, governments intervened, endorsing the securitization of those assets, enhancing their liquidity, and apparently hedging risk through diversification. Moreover, the securitization paradigm minimized the need to use equity in the system since it represented subordinate capital that the wholesaler provided along with the contingent capital provided by the mortgage insurance company to secure the mortgage pool, for which funding came directly from the capital market.

In the late 1980s, rating agencies changed their guidelines to enhance originators, allowing subprime portfolios to be rated as AAA securities and, from then on, private-label markets boomed, and even more in the 1990s (Ashton, 2009). Meanwhile, in the institutional mortgage market, GSEs combined portfolios, which grew exponentially in the early 2000s (accounting for USD 1.3 trillion, representing 23% of the US home mortgage market), a fact that was later criticized by former US Federal Reserve chairman Allan Greenspan, observing a systemic risk should the GSEs not be capable of regulating themselves by monitoring the markets' trends (Andrews, 2008). It was argued that the lack of market discipline and the inability of counterparties in GES transactions to assess risk accurately led firms to rely on the government's endorsement rather than the underlying soundness as financial institutions, later proving that regulators could not depend on the market's discipline to contain systemic risks, or instead, that the market cannot rely on the government to properly regulate.

Financial capital creates multiple mechanisms to operate in institutional settings all over the world, developing instruments such as securities, collat-eralized debt obligations, credit default swaps, or securitized student loans, that is, creating asset-backed security instruments to incorporate assets into the financial circuits. These mechanisms are hindering urban planning in cities since they intervene in the production of spatial configurations affecting the living conditions of millions of citizens. For instance, once a mortgage is sold as a security in the financial market, the solvency of the borrower becomes irrelevant to the system, later resulting in the reproduc-tion of large, vacant low-income housing developments in the periphery of Latin American cities, or in the displacement of tenants from rental housing in the central districts of the city, as the source of profit has shifted from the monthly payment of the mortgage to the sale of the mortgages in the finan-cial market. Moreover, the subprime crisis had multiple consequences for the region, since it "infected Latin America after the second half of 2008 through foreign direct investment flows, decreasing remittances and a sig-nificant drop in international tourism" (Marichal, 2010, p. 312).

A key ingredient in the crisis were the *credit default swaps* (CDSs), which were used as instruments to obtain a profit from predicting failure, and one of the main escalating factors of the subprime crisis. These CDSs are con-sidered part of the shadow banking system as investors profited largely from the crisis, which at the time accounted for two-thirds of all banking transactions. Swaps were sold as insurance to hedge against risk, but as they were classified as derivatives, there was no requirement that they be backed by capital reserves, nor were they subject to the kind of regulation that is required of insurers for similar protection.[7] At some point in 2007, investors attempted to cash in their CDSs when they started to notice the quickening pace of foreclosures, which had already started to devaluate mortgages and CDOs. When housing prices plummeted, mortgages lost correlation with the value of the asset that served as their collateral, and by 2010 it was estimated that around 7 million US households had their homes foreclosed, leaving disastrous social and economic consequences, especially for lower-income citizens (Sassen, 2014, p. 128).

The impact of collateralized debt obligations (CDOs)

> In the first decade of the twenty-first century, a previously obscure financial product called the collateralized debt obligation, or CDO, transformed the mortgage market by creating a new source of demand for the lower-rated tranches of mortgage-backed securities.
>
> (Financial Crisis Inquiry Commission, 2011, p. 127)

The Great Recession[8] is one of the major crises of contemporary globalization affecting capital markets all over the world, through a chain reaction that extended to the highlighted complex system of interdependent financial products (MBSs, CDOs, CDSs, etc.). As a consequence of the unprecedented rate of economic growth that preceded it, the recession revealed the unregulated global finance practices that underlie that growth as well as reminding us once more of the contradictions inherent to capitalism, producing systemic crisis through regeneration processes by the so-called mechanisms of *creative destruction* (Schumpeter, 1942). The Great Recession highlighted key features of late capitalism, such as the structural interdependence of the economy, the continuing expansion of credit, the withdrawal of regulatory agencies, the rise of shadow finance, the reckless relation toward risk, the speculation greed, as well as the unregulated growth of financial markets, inequalities, and disparities in the broader sense. But it is also important to remember that this crisis produced a strong decline in GDP in many countries, directly affecting the lives of millions of low- and middle-class citizens who lost their jobs and houses, and causing growing unemployment, social and political upheaval, and further socioeconomic inequality and dispossession.

Crises are the result of endogenous and exogenous factors inherent to capitalism and can be considered structural in nature, and yet, the actual events are usually framed as conjunctural since specific factors have to converge at some point, usually questioning the management procedures to contain the crisis as well as the role and responsibility of the different actors involved. As the *Final Report of the National Commission on the Causes of the Financial and Economic Crisis in the United States* suggests, the subprime crisis could have been prevented if adequate regulation of financial institutions had been put in place: mechanisms for corporate governance, transparency, and the effective management of risk in financial operations, as well as the enforcement of ethical standards.

Which brings us to a second underregulated financial instrument that led to the crisis, and also reshaped real estate investment, the collateralized debt obligation (CDO). CDOs were first engineered as a credit risk transfer device that would allow investment banks to repackage loans and mortgages into tranches,[9] originating value from the amortization of the corresponding debt (Fabozzi et al., 2016). One of the innovations of capital market instruments was their use of credit derivatives to transfer risk, especially through credit default swaps (CDSs), which enabled mechanisms of synthetic securitization.[10] However, this variant added risk to the banks' trading books, exposing them – as well as investment banks and conduit investors – to concentrated risks and losses in events of crisis or fraud. The use of credit derivatives technology to transfer asset risks and cash flows permitted the creation of *synthetic* CDOs, which had three-quarters of the market, having significant advantages over *cash* CDOs since they had an arbitrage advantage of the equity tranche, among other features, that permitted risk transfer, improving the credit rating and facilitating the diversification of the portfolio.

Securitization is a subset of structured finance that reduces borrowing costs, permits the transfer of the assets' risks and liabilities, and reduces the size of an institution's balance sheets as well as profit from regulatory capital arbitrage and tax management,[11] even when the efficient market hypothesis asserts that information drives prices down, forcing rates to converge, thus eliminating the arbitrage practices. CDOs are backed by portfolios of assets that assemble a combination of MBSs, loans, bonds, securitized receivables, ABSs, tranches of the CDOs, and credit derivatives associated with the former. Normally CDOs rely on special-purpose vehicles (SPVs) to buy the portfolio of assets and issue tranches of debt and equity.

There is an important difference between financial guarantees and credit derivative contracts, since the buyer of such derivatives does not have to own the underlying security or even suffer a loss to claim protection. CDOs can securitize any potential future stream of payments or future value, ranging between different asset classes of residential mortgage loans, including single-family, multifamily, condominium, cooperative, and commercial mortgage loans, but also consumer receivables (such as auto leases, auto loans, home equity loans, and credit card receivables) and student loans, which belong to their own category. Also, CDOs are classified as synthetic (credit derivatives), cash, or a combination of both, which may be backed by CDSs.

A special mention has to be made of *special-purpose entities* (SPEs), also referred to as *special-purpose vehicles* (SPVs), which are either a trust or a company. The SPEs warehouse the asset risk whether by purchasing the assets or in synthetic form, and had been instrumental for structured finance. Such entities can be onshore or offshore, usually off balance sheet, private in nature, and bankruptcy remote, serving legal and illegal ends.[12] This statement is significant, since these instruments have been extensively used for money laundering, to misstate earnings through wash trades, embezzlements, and concealing losses (Tavakoli, 2008). Also, since financial assets are not actually sold as products, their securitization does not get true sale treatment for accounting purposes, therefore reducing regulatory capital according to corresponding accounting principles. Along with the SPE and the SPV, there is a third variant, *special-purpose corporations* (SPCs), also known as shell companies, which have been associated with offshore tax havens and whose ownership structure is undisclosed, often used for tax avoidance, which is legal, unlike tax evasion, which is a felony.

CDOs were considered as a new growth engine for the global security business prior to the 2008 crisis. A CDO will typically pool between 20 and 500 loans or bonds, in contrast to traditional ABSs that assembled loans pools of 500 to 100,000 loans. These instruments may be classified according to the assets' acquisition (cash, synthetic, hybrid), what it held (high-yield, investment grade, emerging market), purpose (balance sheets, arbitrage), leverage structure (cash flows, market, value), and asset ramping (significant increase in the level of output of a company's products or services). Regarding the first class, cash CDOs were funded by an originator or with funding raised from the market, while in the case of synthetic CDOs, credit derivative deals provided protection against the assets, without having to acquire them as such.

Usually, CDOs would raise sufficient cash funding from investors to secure AAA rating on the *senior-most* of their securities, used as a credit enhancement to absorb the risks of the synthetic liabilities of such obligations. To this effect, hedge funds were instrumental for structured finance products where major investments were made in *tranches* of CDOs, taking higher risks with leverage, funding, and the potential of high returns. A key aspect was that hedge funds managed to earn a substantial fee to risk investors' money, getting paid when winning bets, but were not affected in the case of losses.

In the early 2000s, the originators of CDOs were their own dominant buyers, and yet they managed to influence prices in the securities market, which led to funding MBSs and later using them as collateral for the CDOs. At this time, Europe accounted for the generation of nearly 43% of the global CDO issuances, in the process contributing significantly to positioning them as a major instrument of structured finance. In order to understand the rapid development of the CDO market, the role of Alan Greenspan, as the chair of the Federal Reserve of the United States (1987–2006), has to be highlighted, since he regarded risk transfers by credit derivatives and CDOs as central to maintaining the health of the global financial system, as stated in a major conference address:

> As is generally acknowledged, the development of credit derivatives has contributed to the stability of the banking system … They are concerned that banks' efforts to lay off risk using credit derivatives may be creating concentrations of risk outside the banking system that could prove a threat to financial stability.
>
> (Federal Reserve Bank of Chicago's 41st Annual Conference on Bank Structure, Chicago, Illinois, May 5, 2005)

The main innovation that CDOs brought to structured finance was the integration of the MBSs' low-grade investment tranches (largely BBB and A) into new securities. CDOs placed these tranches among risk-averse investors to be paid first, while risk-seeking investors were paid last. The rationale behind this procedure was that diversifying the quality of those securities would be beneficial since mortgages had little chance to default all at once, so risk was tempered. However, the growth of CDOs was exponential, mainly due to the fact that "between 2003 and 2007 as house prices rose 27% nationally and [US] $4 trillion in mortgage-backed securities were created, Wall Street issued nearly [US] $700 billion in CDOs that included mortgage-backed securities as collateral" (Financial Crisis Inquiry Commission, 2011, p. 129). However, theory proved wrong and borrowers defaulted in large numbers, and since correlation was high (performed similarly), an avalanche of defaults took place. There are several elements that contributed to the exponential growth and wide spread of the crisis, in which CDOs played a major role, such as the following.

A regular practice was for those constructing the CDO to include portions of other CDOs, mostly in mezzanine tranches – up to 15%, but the so-called *CDOs squared* included as much as 80–100% from other CDOs. A further development in synthetic CDOs was that instead of containing tranches of mortgage-backed securities or even tranches of other CDOs, they would contain credit default swaps (CDSs), that is, bets against borrowers paying their mortgages, not financing any home purchase. These riskier instruments guaranteed the reimbursement of losses on the tranches in exchange for premium-like payments, issued by insurance companies such as American International Group (AIG), which made CDO investment more attractive, but seriously exposing the CDSs in the case of significant losses. In order to estimate the risk exposure, insurance companies such as AIG examined different length choices to cover the value of insurance liabilities: "As part of its asset/ liability matching discipline, the company conducts detailed computer simulations that model its fixed-rate assets and liabilities under commonly used stress-test interest-rate scenarios" (AIG, 2002).

Another potential ingredient for financing CDOs was *leverage*, that is, investing borrowed money through structured investment vehicles rated as AAA securities. These instruments could be leveraged up to 14:1 (that is, holding $14 in assets for every dollar invested), so the operation could be financed with debt and little or no capital backing. To take things even further, CDOs could be constructed with other CDOs serving as collateral during the origination process, furthering leverage with such practices. Therefore, the equilibrium of the structured financial system relied on the stability of the market value of properties whose mortgages were the basis of the securities, as well as the ability of creditors to meet their monthly payments. Collateral debt obligations' management fees were paid according to the dollar amount of assets and sometimes depending on performance, ranging from USD 750,000 to USD 1.5 million for a $500 million deal.

One of the core elements in this system were rating agencies (Moody's, Standard and Poor's, Fitch Ratings) that evaluated the probability of default of the securities in the CDOs, as well as estimating the correlation between mortgages, that is, the possibility of simultaneous defaults among securities. The agencies used different modeling approaches and they could even choose different stress scenarios when evaluating cash flows with different tests and restrictions. Not surprisingly, the boom in structured finance and the revenue from ratings coincided with the housing bubble prior to the subprime crisis. The rating agencies' fees were typically paid between USD 250,000 and USD 500,000 for a standard CDO rating, whose appraisal provided basic guidelines on the collateral and structure of the security (size and return of the various tranches), AAA rating being a crucial marker signifying the product was a sound investment. Along with this, specific mechanisms were put in place that fueled the housing bubble, such as the CDO assemblage of riskier tranches of the mortgage-backed securities, the inclusion of credit default swaps, synthetic CDOs, and asset-backed commercial paper.

Between 1997 and 2006 housing prices in the United States increased 152%, and the Securities and Exchange Commission (SEC) failed to provide oversight, including requiring adequate holding of capital and liquidity, of the five major investment banks. According to the commission, this lack of oversight was partly responsible for the estimated between 8 and 13 million foreclosures in the United States, depending on who you ask. At the time of the crash, home values plummeted to the point where borrowers had negative equity in the home, that is, their debt was greater than the home's value, which – along with the economic volatility – raised monthly mortgage payments disproportionately and created financial hardships for debtors. Another contributing factor was that the US Federal Reserve increased the demand for risky investments by keeping interest rates too low for an extended period of time. According to Marazzi (2009, pp. 32–33), many saw the financialization of the economy as a way of preventing further devaluation of the rate of return of capital and enabling the recovery of its previous margins of profit.

The Great Recession brought US housing prices down 32% in 2006 and yet by 2012 they had already rebounded (Walks, 2010),[13] resulting in mechanisms of dispossession, somehow reproducing the aftermath of the Great Crash of 1929. Among the causes of the subprime crisis, it has been claimed that countries such as China, emerging economies in Southeast Asia, and oil-producing nations accumulated capital surpluses that entered the United States and European financial systems, causing interest rates to decline, furthering investments in high-risk mortgages and extending this trend to the United Kingdom, Ireland, Italy, Spain, and Portugal (Ito, 2008).

With no effective controls and constant demand from CDO issuers, originators had no incentive to restrict lending to sustainable mortgages, and homebuyers took advantage of this, upgrading their expectations to more expensive properties because they could refinance their previous mortgage. Meanwhile rating agencies continued earning substantial commissions by giving these CDOs AAA ratings. Even when home prices kept escalating, demand was not affected as a result of the creation of interest-only adjustable-rate mortgages or by alternative paying options that reduced down payments or initial monthly payments that did not even cover the cost of interest. At the peak of the bubble, mortgage-underwriting standards were reduced to little or no documentation, allowing the already mentioned *NINJA* loans: no job, no income, and no assets. These, too, were bundled into CDOs and given AAA ratings.

Other mechanisms also proved to be critical for furthering the crisis: the nature of CDOs meant that brokers were not responsible or even negatively affected by badly performing mortgages; the complexity of financial terms made it difficult for borrowers to understand the potential risk if home values declined, but also the government's endorsement of legal but questionable lending cast doubts on the system as a whole. Moreover, the securitization and collateralization mechanisms increased the vulnerability of the financial environment surrounding mortgage markets, where insufficient regulation played its part along with negligence, with both factors increasing the devastating effects of the crisis: "The key differences in this case were leverage and risk concentration. Highly correlated housing risk was concentrated in large and highly leveraged financial institutions in the United States and much of Europe" (Financial Crisis Inquiry Commission, 2011, p. 427).

Therefore, concentration of highly correlated risk, insufficient capital investment related to debt (some institutions were leveraged 35:1 or higher), and poor risk management systems led some institutions to go bankrupt while handling toxic assets. Some questions have been raised regarding the lack of adequate regulation by the Securities and Exchange Commission (SEC) before the crisis, arguing that countries with stricter regulatory regimes also failed and had to be bailed out, including the United Kingdom, Germany, Iceland, Belgium, Netherlands, France, Canada, Spain, Switzerland, and Denmark. In those cases, loans were usually made by federally regulated lenders and yet risky investments in structured finance were made because regulations allowed them to do so. Finally, federal authorities argued in the US that the State's bailout prevented the collapse of the global financial system comparable to the Great Depression of 1929, which seemed an all too convenient explanation facing the unfolding crisis.

Credit default swaps (CDSs)

Credit default swaps (CDSs) are defined as agreements between two counterparties to exchange periodic fees (also known as *spreads*), in return for a payment contingent on a credit event. The two counterparties are known as a protection *seller* and a protection *buyer*. The latter makes periodic payments (premiums) to the former, who is obligated to pay compensation in case the asset experiences a credit event, such as bankruptcy or liquidation. A key feature of CDSs is not requiring investors to hold the position that needs protection – as when owning an asset. While the purchase of CDSs requires quarterly spread payments to the protection seller, in the case of a credit event the spread is calculated according to the value of the underlying asset, as a percentage of its value according to the probability of the credit event and the recovery rate.[14] Sometimes, an investor holding a long bond position could be exposed to the probability of a credit event, for which one could purchase a CDS as a protection buyer with which the investor's portfolio could remain neutral to changes. Likewise, an investor holding a short bond position could use a CDS as a protection seller to cover exposure to the unlikelihood of a credit event.

Among credit default swaps, there are the *single-name* CDSs, where their cash flow and value resides on the credit quality of a single entity (sovereign, corporation, or municipality); *multi-name* CDSs based on the credit risk of more than one underlying reference name, which can be classified as either *portfolio* or *basket* CDSs and are compensated for credit-related losses on all or any reference entities in the portfolio; the *index* CDSs' cash flows and values are tied to an index or portfolio of multiple references entities that meet specific criteria of the index provider; while *tranched index* CDSs enable credit protection purchases on specific tranches of the underlying index.

Asset-backed credit default swaps rely on the credit quality of a specific underlying security or asset, and their value and cash flow depend on a single specific reference asset or security, using a special-purpose entity (SPE) which can be a trust or a corporation established to facilitate structured financial transactions. This kind of CDS was particularly affected during the subprime crisis, and it has been argued that there was an inherent *design* in such products; nonetheless, it is likely that this instrument could come back to the global credit derivatives marketplace in the near future.[15]

Securitization mechanisms emerge from two major motivations: first, there are asset originators who are looking to sell credit-sensitive financial products while reducing credit exposure to the underlying pool of assets; second is its use for *arbitrage* securitization that comprises credit protection sales of the underlying assets by a collateral manager who collects a fee. Asset-backed securities have been associated with complex financial structures, but, in fact, the features embedded in the contracts (such as cash-flow waterfalls) vary greatly, and have to be evaluated on a deal-by-deal basis. However, CDS operations can conceal important information from investors since they are entitled to transfer an asset credit event risk to a CDO without notifying the borrower of this transaction and the risk involved.

Warren Buffet (2003, p. 15) famously called derivatives *financial weapons of mass destruction*, anticipating the role that such products would play in exacerbating the financial subprime crisis of 2008, and in particular AIG's spectacular credit default swap losses that led to its bailout. However, Alan Greenspan's address to the Federal Reserve Bank of Chicago a couple of years later was more sympathetic to the role of derivatives in managing credits and risks more effectively:

> In particular, the largest banks have found single-name credit default swaps a highly attractive mechanism for reducing exposure concentrations in their loan books while allowing them to meet the needs of their largest corporate customers [and yet], some observers argue that what is good for the banking system may not be good for the financial system as a whole. They are concerned that banks' efforts to lay off risk using credit derivatives may be creating concentrations of risk outside the banking system that could prove a threat to financial stability.
>
> (Kothari, 2006, p. 420)

This quote implies that the US Federal Reserve was aware of the risks involved in the use of derivatives and yet did not support adding regulations to capital markets that had already reached USD 62 trillion by 2007, and where the USD 800 billion subprime mortgage market threatened to bring down the financial system (Sassen, 2012, p. 76). Even if the banks' bailout prevented the imminent collapse of the US economy, the impact of the subprime crisis led to 2.2 million foreclosures of modest-revenue households in the USA, leading to mechanisms of capital extraction and accumulation by dispossession (Harvey & Rivera, 2010). Allegedly, CDSs were originated as financial derivatives instruments to hedge risk by transferring it to entities willing to bear it, enabling financial institutions to make loans they might otherwise be unable to make, in the process revealing useful information about credit risks in their prices. However, CDSs enabled unsustainable credit levels, excessive risk-taking by financial institutions, as well as market manipulations, where hedge funds and investors bet against CDSs, creating a network that enabled short-term gains and increased the possibility of systemic risk. A next cycle ensued and CDSs recovered again, reaching USD 17 billion by 1985, followed by a Tax Reform Act in 1986 that included a modernization amendment that enabled considerable growth of CDS-related assets, providing greater flexibility management as well as a better tax environment.

When the 2008 subprime bubble burst, CDS owners wanted to cash in their securities as insurance policies, only to find out that there had been no requirement that they be backed by collateral. This was possible because these securities were considered to be derivatives, in the process eliminating the capital reserves required of actual insurance (more about this below). Derivatives had been used to hedge against risk but also to speculate on changes in rates, prices, and potential defaults on debts. However, the escalation of risk was produced by uncontrolled leverage,

lack of transparency, collateral requirements, and the correlation and concentration of risks in the market. Facing this scenario, CDSs were sold
to investors as protection against default or the decline in the value of
mortgage-related securities, which were backed by risky loans, in the process expanding the market and fueling the housing bubble. In the first
decade of the twenty-first century, derivatives accounted for nearly USD
630 trillion, equivalent to 14 times the value of global GDP, for which the
financialization of the economies became a global phenomenon (Sassen,
2012, p. 81).

Before the Great Recession, an influential paper was published by
former chief economist of the International Monetary Fund, Raghuran
Rajan (2005), warning of the potentially disastrous consequences of an
unstable financial system that could prove incapable of containing fallout
from the bypassing of capital requirements for financial institutions
through hedging them with derivatives. At the time, Alan Greenspan and
Laurence Summers acknowledged that CDSs were problematic and they
understood, maybe too late in 2008, that "the 100-fold growth between
2000 and 2008 was unprecedented and that by that time ... our regulatory framework with respect to derivatives was manifestly inadequate"
(Financial Crisis Inquiry Commission, 2010).

Hedge funds are investment firms that have less thorough regulations
and are capable of trading assets with fewer restrictions, engaging in
high-risk practices but also generating high returns to investors through
different arbitraging mechanisms, including exploiting variations in estimated value between different market segments. These funds would usually acquire CDSs to take offsetting positions in different tranches of the
same CDO securities, even betting against the housing market. By the
mid-2000s, hedge funds started shorting the most vulnerable tranches
of the mortgage-related bonds, relatively cheaply and with considerable leverage, after realizing that home prices were escalating without a
corresponding rise of income or wages. CDSs introduced new risks and
leverage to the financial system while their investors did not take out a
single loan or purchase an actual mortgage, simply by making a side-bet
on the risk of default by others, without exposing their own capital.

Another key feature of CDSs is that prior to the US financial crisis
they were treated as deregulated over-the-counter (OTC) derivatives,
which meant they were exempt from regulations applying to insurance
products that protect investors against default of the underlying assets
through backup capital to cover their exposure. Before the crisis, CDS
risk was concentrated in half a dozen major financial institutions specialized in derivatives, including AIG Financial Products, JPMorgan Chase,
Citigroup, Bank of America, Wachovia, and HSBC, which were the more
affected institutions during the financial collapse. In particular, AIG
Financial Products had been selling CDSs to European banks in a variety
of financial derivatives in exchange for a stream of premium-like payments

that allowed banks to hold less capital against their assets (Financial Crisis Inquiry Commission, 2011, p. 350). Also, CDSs played a major role in the creation of synthetic CDOs, which were side-bets on the performance of mortgage-related securities and, at the time of the collapse, amplified the impact of the crash of the housing bubble by allowing multiple bets on the same security, augmenting with it the resulting spread of the crisis through the financial system. Even if it is a contested point, CDSs have been credited with a number of potential correlations that would make it appear that they have led to structural risk across financial institutions that may have served as a mechanism of transmission in the case of major crisis, in which correlations and covariance tend to exhibit elevated volatility during times of market-wide uncertainty.

As a result of the crisis, the heads of state of the G20 nations expressed their commitment to rebuild a "fundamentally stronger financial system than existed prior to the crisis" (G20 Pittsburgh Declaration, 2009, p. 8). Following the Great Recession, major changes to the global financial regulatory framework took place, which made the CDS market decline.[16] Among these adjustments were the revision of the *Basel Accord* that increased the capital costs of trading single-name CDSs and a European ban on short-selling CDSs without owning an obligation of the reference entity. Moreover, the relatively low default rates on corporate debt in the following years also rendered less attractive the use of CDSs for risk-hedging transfer purposes.

In the case of the United States, Congress appointed a commission to investigate the causes of the financial crisis of 2008, who concluded that

> AIG failed and was rescued by the government primarily because its enormous sales of credit default swaps were made without putting up initial collateral, setting aside capital reserves, or hedging its exposure – a profound failure in corporate governance, particularly its risk management practices
>
> (Financial Crisis Inquiry Commission, 2011, p. 352)

Moreover, the company's procedures and interconnections with other large financial institutions created systemic risks endangering the US economy for which a bailout of more than USD 180 billion had to be used for its rescue and, in fact, investors were totally bailed out. Later, regulators recognized that CDOs and CDSs could concentrate instead of diversifying risks.

There is an implicit rationale behind the widespread demand for CDSs as credit-risk transfer instruments, since banks that hedged risks from their borrowers' defaults were incentivized to originate larger and riskier loans, which developed into moral hazards and induced institutions to engage in unreliable lending practices as well as contributing to leverage financial markets. Another moral liability ensued when banks hedged their credit risk exposure as rigorously as they could, shifting monitoring responsibilities to credit protection sellers with less expertise than financial institutions already had. An even more predatory practice turned out to be *negative economic interest*, in which creditors hedged their positions, buying a substantial amount of debt and later purchasing protection against default, for which they could be tempted to direct the underlying asset-related company to bankruptcy, where "By purchasing a material amount of a firm's debt in conjunction with a disproportionately large number of CDS contracts, rapacious lenders (mostly hedge funds) can render bankruptcy more attractive than solvency" (The Economist, 2009).

Real estate investment trusts (REITs)

Real estate investment trusts (REITs) are closed-end investment companies, many with publicly traded stock, that enable capital flows from investors to the real estate market. As real-estate-specific financial intermediaries, REITs were approved in the early 1960s in the US by federal legislation that provided tax exemptions to entities satisfying specific requirements. These requirements include having 75% of a REIT's gross income coming from rents, mortgages, government securities, and real estate sales, and at least 95% of its taxable income distributed among its shareholders annually (Gotham, 2006). With these provisions, REITs turned into an asset class, where real estate not only became a sector for large investors but opened opportunities for individual stockholders who wanted to pool their resources to invest in large-scale profitable commercial real estate modeled after mutual funds, in which a portfolio of buildings could be assembled and traded on the stock exchange market.

At present, there are about 225 publicly traded REITs in the US covering the office, industrial, apartment, shopping center, regional mall, hotel, health care, and specialty property sectors. They are registered with the Securities and Exchange Commission (SEC) and trade on the New York Stock Exchange with a combined market capitalization of more than USD 1 trillion (Cepni et al., 2021). REITs comply with the Modern Portfolio Theory (MPT), which is based on the idea of mixing different investments such as stocks and bonds in a portfolio to improve the expected return and lower the volatility risk over time,[17] as well as the correlation of returns (Markowitz, 1959).[18] A key feature is that REITs tend to be less sensitive to changes in the interest rate environment than both bonds and the broader stock market. According to correlation patterns and historical data, returns from REITs vary during different interest rate periods, but for the most part they have shown a positive correlation during increasing interest rates. For example, the correlation coefficient between government bonds and the S&P 500 was measured at 0.42, as compared to the correlation coefficient of total return from the National Association of Real Estate Investment Trusts (NAREIT) Equity Index[19] (a REIT-based index) and government bonds, which was 0.24, suggesting that, contrary to popular belief, REITs are less sensitive to changes in long-term interest rates than the broader equity market. Along these lines, NAREIT kept track of REITs' performance, categorizing them as *equity debts* REITs, which produced their income through real estate (45% of the total); *mortgage* REITs, consisting of debt instruments secured by mortgages (46% of the total); and *hybrid* REITs, combining direct ownership of real estate and mortgage debt (9% of the total).

The major financial innovation that REITs brought to the table was the possibility of transforming real estate assets and mortgages into financial securities, granting low transaction costs of otherwise illiquid assets. Indeed, while some of the less positive attributes of real estate have been its lack of liquidity, immobility, location, performance appraisal, and liability to externalities linked to the local realm, all of these were mostly solved with the REIT configuration. The hybrid condition of these distinctive financial instruments benefits from the economic features of the underlying asset (real estate), and yet handles the volatility of the stock market. Through this configuration, investors have been able to participate in a broad range of property sectors as well as in various geographical locations. This flexibility positions REITs as key players within the institutional investment arena, attracting retirement fund, insurance company, and commercial bank investments.

In the case of *equity* REITs, cash flows out of rent paid to the building owner are considered as holding low risk when matching future abilities, as well as securing consistent and predictable cash flows, which perform better than financial assets in an inflationary environment. In fact, many portfolios tend to include real estate assets to hedge the effects of inflation. Also, real estate is considered to have low correlation to other financial assets such as bonds or stocks, enhancing its returns and lowering risk in their portfolio (it is suggested that investing 5–20% in real estate significantly increases the total return and lowers risk exposure). Other benefits come from the depreciation of the value of real estate assets, which is very convenient for tax accounting purposes since generally it is greater than the actual economic life of the property, and, nonetheless, a well-maintained property in a good location may increase its value over time at a similar rate of inflation. Moreover, REITs can also engage in partnerships to allocate higher levels of cash flows to tax-exempt investors that allow incremental total return to be enhanced.

As discussed in Chapter 2, real estate has particular features distinct from regular financial assets regarding its value, such as location, physical attributes, and externalities. Among them are zoning provisions, property rights, and land-use control, which correspond to the *highest and best use value*, but also should serve the public interest. Zoning is a physical planning technique that regulates land uses, while considering economic activities, adjacent uses, transport and communication facilities, environ-mental protection easements, and entitlements.[20] Since these ordinances are binding, some planning authorities may not permit real estate developments in certain areas and promote instead urban refill operations or the redevelopment of existing sites and properties.

Real estate can hedge exposure to inflation whenever there is a period of escalating property prices. However, financial capital also tends to flow into the market creating a speculative bubble, which, combined with inadequate regulatory oversight and leveraging mechanisms, may heighten the scale of an upcoming crisis, just as the market crashes in the mid-1980s as well as the subprime crisis of 2008 came close to bringing about the collapse of the entire US savings and loan systems. As Minsky (1982) stressed, economic crises are cyclical and represent major social challenges since it takes decades for the economy to absorb the excess supply of real estate and the financial losses of debtors and investors and reestablish the living conditions of the population.

During their first decade of existence, the assets held by REITs grew from under USD 1 billion in 1968 to USD 21 billion by 1975, mainly from mortgage-related investments, and yet this market suffered its first major crisis between 1973 and 1974, derived from poor underwriting and rising interest rates that ended up in bankruptcy and liquidation of sev-eral firms.[21] Nonetheless, a new crisis impacted REITs between the 1980s and 1990s, known as the Savings and Loan (S&L) Crisis, when savings and thrift institutions originated mortgages and other personal loans to individuals while interest rates started to escalate swiftly. This produced a recession which, along with regulation shortfalls and Ponzi schemes, brought these banks to bankruptcy and led to a nationwide recession in the real estate industry, requiring the Federal Deposit Insurance Corporation to undertake a massive bailout of the S&L system. After the crisis, REITs started a new cycle in which investors were more confident to use them as efficient conduits to access capital from the public marketplace. A major attractor for investing in real estate has been the consistent and predict-able cash flow from rents, its hedging features against inflation, and their ability to diversify portfolios, rendering it relatively independent from the stock exchange market. Publicly traded REITs in the United States hold USD 3 trillion in real estate assets, and account for an average of USD 9 billion in trading volume (as of May 2020).

There are different strategies for integrating REITs in a portfolio, such as owning individual REITs, which allows a direct investor to make its own decisions. Another approach is directing a percentage of a diversi-fied portfolio to be managed by a specialized professional in management of real estate accounts or in real estate mutual funds using *unit investment trusts*, which are self-liquidating pools with a determined time span; and finally, the *exchange traded funds* (ETFs) that follow a market index to engage in stock-like trading, such as short selling or buying on margin.

Financial markets that include real estate assets go through cyclical variations in time, where they may attract new marginal capital that will appreciate such assets to higher levels – in which market bubbles are not rare – or conversely, they can lose value swiftly, causing disruption and correlated fallouts. It is also expected that after a crisis, real estate values eventually recover to their market-level value, for which they are considered a long-term investment. On the one hand, real estate markets have particular features regarding supply and demand mechanisms, such as a limited supply of available vacant space for lease and sublease, new spaces under construction, and future vacant spaces, while on the other hand, the demand for space relates to new business creation, expansion, downsizing, or closing operations, the total reflecting the marginal demand for space within the territory.

Within the real estate arena, the residential housing market has led the way into and out of economic recessions, while commercial real estate markets have usually followed the overall economy, yet the market developments have been highly local in nature, being far more decisive for the valuation of real estate property than the general economic trends. *Residential* REITs represent around 18% of the total capitalization of the NAREIT Equity Index, and produce an average annual return of 16.3% (the highest of any REIT sector), although the volatility of this segment as measured by the standard deviation of returns is 16%, which is also significant. The *office* REITs have changed considerably in the last half century, where suburban office space exceeded downtown office space, growing twice the rate of central business districts (CBDs).

This points to a tendency toward the suburbanization of residence, first starting in the 1950s after suburbanites followed highways to single-family homes and shopping amenities, and where jobs moved to the suburbs, creating metropolitan areas. Moreover, public transport systems, zoning provisions, and existing road networks may have reinforced suburban dynamics of decentralization. The US has the highest ratio of retail space per person anywhere in the world, and retail operators are part of major trade networks and organizations with access to public credit markets for which they are graded by major credit rating agencies (Moody's, Standard & Poor's, Fitch Ratings) as Class-A retail properties. In this regard, *retail* REITs assemble super-regional malls, regional malls, shopping centers, power centers and big-box retail spaces, outlet centers, as well as neighborhood and community centers.

It is important to highlight that property size and age categorize retail complexes, since they tend to follow a downward spiral of decline into obsolescence leading to the abandonment of the facility by the retail tenants. While retail properties' appraisal depends on the range as well as on the market location, they are subject to externalities such as escalating competition (as when a large tenant such as Walmart opens a store near a grocery store) and functional business models or new trends in the retail sector (such as drugstores preferring to be freestanding units outside a mall); *retail* REITs represent 20% of the NAREIT and represent 30% of all investment-grade commercial real estate.

Within the commercial real estate market, *hotel* REITs have historically been the most volatile sector, following the economic overall trends and with a standard deviation of returns of 41.1. *Industrial* REITs represent about 8% of the market equity index, comprising warehouses, manufacturing mixed-use and special purpose uses (such as loft spaces); *self-storage* REITs represent 4% of the capitalization of the NAREIT Equity Index within an aggregate capitalization in excess of USD 6 billion; and among *specialties* REITs are golf courses, movie theaters, prisons and gas stations, automobile dealerships, etc. *Mortgage* REITs represent 3% of the NAREIT, originating or owning loans and other obligations that are securitized by real estate as collateral, which do not meet the criteria to be packaged as MBSs, also known as nonconforming loans, specialized in adjustable-rate mortgages that respond to a predetermined index, which were considered as highly risky assets as the interest rate could suddenly increase the debt. Finally, *health care* REITs represent 5% of the NAREIT Equity Index, with a market capitalization of nearly USD 7 billion, while health care makes up about 17.7% of US GDP. Approximately half of health care REITs' investments are directed towards the nursing home property sector. Other emerging uses are acute care hospitals, rehabilitation centers, medical facilities, office building clinics, and psychiatric and substance abuse hospitals, among others.

At the beginning of the twenty-first century, REITs were the best performing financial investments, with an annual total return of 11.1%, and their aggregate equity market capitalization increased from USD 8.7 billion in the early 1990s to USD 539 billion in 2015, for a compound annual growth rate of 34%, and even if volatility increased dramatically bringing about greater risks, it also allowed high returns. In addition to assets held by public shareholders, most REITs are managed by private, non-traded *operating partnership* (OP) units, which are similar to shares of common stock as they represent a percentage of the ownership of a REIT but which are not publicly traded.

REITs have provided attractive dividend yields that generally increase faster than inflation, making them an effective hedge while also serving as a diversification tool. On the one hand, equity REITs rent their properties to tenants according to leases, which usually protect their margins from inflation, and whose operations are subject to daily scrutiny from the analyst communities, specialized indexes, and equity research companies. On the other hand, commercial real estate depends on the role of the State to operate and remake real estate markets, enabling financial instruments such as REITs through federal legislation, but later exporting them to countries such as the Netherlands (1969), Australia (1971), Brazil (1993), Spain (2009), and Ireland (2013). In these last two countries REITs were visualized as possible investment vehicles to sell large portfolios of real estate properties acquired by the State after bailing out the banks during an acute financial crisis a few years earlier (Aalbers, 2019a, p. 9).

Financialization creates perverse investment dynamics in the sense that speculators can keep newly built spaces empty for an extended time, either to generate a fictitious shortage or to capture the increase in value derived from real estate speculation in the short and medium term. In addition, mechanisms of so-called *creative destruction* are generated where historic or strategically located buildings are demolished for the construction of high-rise buildings in their place, adding density and height in commercial avenues or entire districts of the city.

Financialization transforms fixed assets into liquid assets that can be traded on capital markets, via which massive investment necessarily has a spatial impact on cities. This capital fixation, or spatial fix, creates conditions conducive for indefinite reproduction, where pension, mutual, and insurance funds invest in the real estate markets with insufficient international regulations, and tax havens feed private and hedge funds. These activities, combined with the financial derivative market, in which bets are made on the fall in the value of shares or sovereign certificates of a particular country, and the capital derived from money laundering, amount to USD 3 trillion annually.

In the following chapters several case studies in Latin America are examined, showing the way in which financial mechanisms operate in the housing sector, in large infrastructural projects, as well as in major tourist developments. The analysis identifies a major shift toward capital investments in real estate that is producing economic disruptions in the urban realm, resulting in dispossession, segregation, and displacement.

Notes

1 The term commercial real estate generally refers to offices, industrial (factories, R&D facilities, and warehouses), hotels, shopping, etc.).
2 *Call* options provide the right to purchase and *put* options the right to sell an asset.
3 Created in 1938 but became publicly traded in 1968.
4 The secondary mortgage market is a marketplace where home loans and servicing rights are bought and sold between lenders and investors.
5 Tranches are slices of a collection of securities, split up by risk or other characteristics separated in order to be marketable to different investors, carrying different maturities, yields, and degrees of risk and privileges in repayment in case of default.
6 RMBS stands for residential mortgage-backed securities, as opposed to CMBS, which stands for commercial mortgage-backed securities and are used for commercial properties.

7 The official narrative of the Great Recession blamed households for defaulting on their mortgages, while credit default swap owners who were trying to cash in their securities at the same time were also a major part of the problem.

8 The term Great Recession applies to the financial crisis lasting from 2007 to 2009, which began when the US housing market bubble burst, and many mortgage-backed securities (MBSs) and derivatives lost significant value, causing millions of foreclosures as well as job losses.

9 A *tranche* is a portion of a mortgage-backed security containing a set of mortgages that are equally risky. Mortgage-backed securities are divided into a set of tranches with the riskier tranches offering a higher rate of return. The tranches are sold to different investors based on how much risk they are willing to take in order to receive a higher return.

10 In this kind of securitization a bank buys credit protection on a portfolio of loans from an investor, and unlike in a regular sale transaction, the loans remain on the bank's balance sheet.

11 *Arbitrage* means the opportunity to borrow and lend at two different rates of interest and includes the possibility to simultaneously buy and sell the same security in different marketplaces for a profit with no risk involved.

12 *Bankruptcy remote* describes an entity created to develop, own, and operate a special project while isolating financial risk and minimizing bankruptcy risk, such as a single-purpose entity, a special-purpose vehicle (SPV), or a special-purpose entity (SPE).

13 Although uneven across the regions, particularly hit by foreclosures were California, Nevada, Florida, and Arizona.

14 That is, the percentage loss that asset may experience due to the credit event calculated as 100% of the reference asset's face value minus the recovery rate.

15 For an opposite argument, see Duffie & Thukral, 2012.

16 The Basel Framework is the full set of standards of the Basel Committee on Banking Supervision (BCBS), the primary global standard setter for the prudential regulation of banks.

17 Often referred to as the *standard deviation of return*, it measures the spread of values in a set of data. For investment analysis, the standard deviation is the most commonly used measure of investment volatility over time, where a lower standard deviation helps to moderate portfolio risk, but it also tends to provide lower returns.

18 A *correlation coefficient* is a statistical measure that shows the interdependence of two or more random variables. The number indicates how much of a change in one variable is explained by the change in another. A score of 1.0 is perfect correlation with each variable moving in unison; a score of −1.0 is perfect non-correlation with each variable moving opposite one another.

19 The sector has had a long history of consolidation in which NAREIT has played an important role of monitoring their performance through their index, which dates back to the early 1970s, and included all publicly traded REITs in a relative market on a monthly basis. Later, a special *NAREIT Equity Index* was designed, which excludes mortgages to reflect a pure equity real estate benchmark, a *NAREIT Mortgage Index* also available on a real-time

basis, and a *NAREIT 50 Index* that monitors the 50 largest publicly traded REITs in the US. Besides these indexes, S&P REIT Composite Index covers 75% of the REIT market capitalization, and the Morgan Stanley REIT Index covers tradable real estate markets, among others.

20 *Entitlement* is the legal right granted by state and local real estate zoning authorities to build or improve a parcel of existing real estate, normally unimproved land.

21 At the time, stock prices of all REITs plummeted more than 85% in just one year.

4 Financing housing markets

Housing: financial markets and the State

The financialization of real estate markets has been framed by mainstream economics through abstract, predictive, and synthetic mathematical models, but in order to understand their change, diversity, and significance, considerations have to be made taking a broader approach that understands them as socially situated economic processes. From the urban planning perspective, markets have had a major impact on the physical structure of cities, shaping spatial relations and affecting the quality of life of citizens through their social, economic, and spatial implications. At the core of financial market crises in recent years, real estate has played a crucial role in creating and bursting economic bubbles, as in the case of the Great Recession in the United States (2007–2009), where subprime loans were a major catalyst, although set within other contingent circumstances. Among these developments there were important correlations between the financialization of real estate assets and global trade mechanisms that enhance their potential for expansion, in which the State has had a privileged position for deregulating some procedures while enforcing tighter restrictions on others, enabling the flow of international capital and creating expectations for the continuous growth of markets.

The different forms of appropriation of land have been a central feature for the creation of States, being the material basis in which production takes place and with it, the reproduction of power and wealth. As discussed earlier, land has a use value as well as an exchange value. The latter can be used as collateral for mortgages, giving the lender a contingent right to the property in case of default, assuming that the State protects property rights and guarantees the regulatory framework. In this context, access to adequate housing and shelter has been granted as an economic, cultural, and social right in the Universal Declaration of Human Rights of 1948, for which the expansion of mortgage markets has been considered a vehicle to increase homeownership for low-wage populations. However, in the last two decades the creation and trade of asset-backed securities changed the social housing rationale, since financial capital targeted real estate

DOI: 10.1201/9781003119340-5

markets through political negotiations that reshaped trading operations, processes that were later institutionalized, resulting in a symbiotic, interdependent relation between the market and the State, which prevailed over social aspirations.

This is where the mortgage-based products discussed in the last chapter come into play since they were at the core of the framework that was set up to enable a secondary market to exchange them as securities, categorized according to credit scoring, predictive behavior, and customer profiling (Thomas, 2000). In most countries, the State played a central role in setting up a structure that enabled the creation of financial markets, taking into consideration the global environment and shaping a corresponding set of regulations. The recurrent narrative is that for a few decades, mortgage markets expanded to the point of exceeding demand, fueling a housing bubble that encouraged predatory lending, targeting borrowers into engaging in high-risk loans that later led to home foreclosures as well as the abandonment of millions of housing units (Valenzuela Aguilera & Tsenkova, 2019). Under these conditions, the refinancing of mortgages takes place when house prices increase rapidly and people take larger loans, later resulting in negative equity, default, and foreclosures, in which the over-accumulation of capital comes out of the financial markets, especially through the securitization of mortgage loans. The result of this process has had disastrous social outcomes, where foreclosures affected more than eight million households during the subprime crisis in the United States (Aalbers, 2009, p. 288).

At the turn of the twenty-first century, housing became a keystone for the financial markets, and yet, securitization changed considerably depending on each country's historic development or economic framework, and shaped housing policies in relation to capital investments in the sector, as well as their integration into the global financial system. Housing stock has been considered to hold a secure, fixed value as collateral, while the securitization of such assets allows the standardization of the mechanisms to capture income flows, rents, and other spreads through an institutional framework that guarantees the storage of surplus value in a relatively safe financial environment (Aalbers, 2017).

However, less remarked upon is that an important shift has occurred in which developers and real estate companies that had traditionally influenced urban policies are now being taken over by financial institutions that, by funding certain sectors of the real estate market, have favored particular spatial patterns correlated with the most profitable use of land. These dynamics have made households increasingly dependent on the performance of financial markets – shareholders who invest in real estate assets to store and increase the value of their wealth – rather than demanding a more comprehensive intervention of the State to provide public welfare for the population.

In the last four decades, capital markets have been shaping public policies while states have softened financial regulations, engaging in the structural reforms needed for establishing financial procedures across the government and furthering the transformation of the State itself. These public policy responses have made states interdependent with the interest of capital through the operation of financial markets, beyond the public interest, welfare policies, and comprehensive city planning concerns in the long run. As the State develops market-making capacities, social demands are relegated to secondary importance, where they remain a neglected priority. The securitization of mortgages may not necessarily lead to broader access to housing since the construction of more units has not prevented the escalation of prices, displacement, or gentrification, or even the privatization of the housing stock through the acquisition of foreclosed properties. By combining financial and non-financial mechanisms, financialization has resulted in different outcomes and particular procedures depending on the regulatory conditions and legal frameworks of each country (Fernandez & Aalbers, 2019).

One element of this multifaceted process is how global markets assemble economies with different degrees of development, in which dependency relations are established toward hegemonic nations through trade arrangements, as in the case of eastern countries with the European Union, Brazil toward China, and Mexico in relation to the United States and Canada. In such scenarios, their interdependence favors direct investments by multinational corporations, disassembling their domestic economic infrastructure and instead articulating them to the global trade chains (Fernandez & Aalbers, 2019, p. 4). In the case of emerging economies in Latin America, they may turn into *global devices* (Valenzuela Aguilera, 2013), spatial hubs that attract capital flows from multinational companies located in advanced capitalist countries, which reinforce inequality, polarization, interdependence, and also accentuate contradictions, since only certain parts of the economy (along with their territory) will develop and store wealth, while the vast majority remain as subordinate participants of the global transformation.

The promotion of large infrastructure projects, including housing, is considered to be a high-risk investment even if profits are privatized and losses – as in the case of the subprime crises – tend to be socialized, reducing public authorities' capacity to dictate the terms and conditions of such undertakings, arguing instead on the risks of such enterprises. Nonetheless, multinational development banks such as the World Bank fostered the promotion of large infrastructure projects, attracting global investors to fund projects that met the millennium goals through financial mechanisms. In the case of countries such as Brazil and Mexico, the financialization of housing programs attracted a low-income population, formerly excluded from the financial markets (i.e. *Minha Casa, Minha Vida* program in Brazil) or boosted second residence mortgages for individuals who qualified for institutional loans (by public funds such as INFONAVIT and FOVISSSTE in Mexico), and yet these mortgages resulted in millions of vacant houses.

According to the International Development Bank standards, the consolidation of housing production mechanisms needs a solid market-based system of mortgages along with well-targeted subsidizing mechanisms. In the case of Latin American countries, Chile is considered to have succeeded in issuing bonds to finance social housing, while Mexico and Colombia have engaged in creating a secondary market of mortgage-backed securities. Nevertheless, it is important to mention that in the region 40% of the population work in the informal economy, which limits their access to financial services (Fig. 4.1). For instance, in industrialized countries over 90% of households have access to formal financial services, compared with 60–80% in Chile, 40–60% in Brazil and Colombia, and 20–40% in Argentina and Mexico (Honohan, 2008; Caskey et al., 2006; Solo & Manroth, 2006), while the ratio of financial system deposits to GDP in 2017 accounted for 80.83% in countries like the United States compared to 50.71% in Chile, 30.58% in Mexico, and 23.39% in Colombia (World Bank, 2021).

An advantage for economies with pools of large contractual savings funds (such as pensions and insurance) is that mortgage securities can use these funds for housing mortgage purposes, which allows them to allocate risks and liabilities within the financial system more effectively than other savings schemes. For instance, Chile has used mortgage bonds (*letras de crédito hipotecario*) extensively, which are on-balance-sheet mortgages slightly different from off-balance-sheet mortgage securitizations, also known as RMBSs. A mortgage bond is a senior general obligation issued by a bank, which is backed by a particular pool of residential mortgages

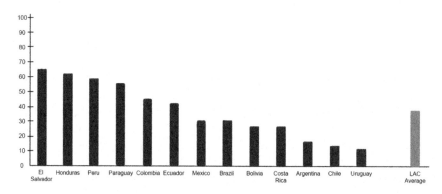

Fig. 4.1 Share of the informal economy in Latin American countries.
Source: OECD et al., 2019.

on its balance sheet and has a collateral pool for the mortgage bond that is separate from the assets of the banks, which are available to meet the claims of the bondholders in case of default. In Colombia a securitization conduit known as Titularizadora Colombiana has securitized nearly 30% of outstanding mortgages since its creation in 2002, selling around half of those bonds to insurance companies, finance institutions, and pension funds. Chile and Mexico are major users of inflation indexation when issuing mortgages, which enables them to provide long-term securities for institutional investors, while regulations were put in place to allow pension funds and insurance companies to allocate investments in the long-term mortgage-based securities market.

Notwithstanding, in Latin American countries different underlying principles prevail regarding housing rights, since a long tradition of social movements are associated with them, even extending them under the right to the city, representing a culture of resistance against foreclosures and evictions which extend those processes for a number of years. This rationale often prevents the establishment of legal frameworks for mortgage bonds and securities that could guarantee foreclosure procedures. In contrast, financial logic demands high-quality lending requirements, the registration and transfer of titles, secure land banks available for development, a clear urbanization standard that developers must meet to be granted the corresponding authorizations, as well as the institutional endorsement for the financial practices of real estate markets. Finally, before turning to housing in individual countries, it is important to make a brief comment about the impact of COVID-19 on the mortgage market, which is ongoing at this time (Fall 2021).

The pandemic has impacted the financial services affecting debt conditions and levels, since mortgages are the single largest source of debt for individual homeowners and has the greatest impact on their finances during crises. However, the recovery of this pandemic may not be uniform as institutions may perform differentially. In the case of mortgage entities in Latin America, there were financial problems before COVID-19 and now it may be wise to adopt the option of 3–18-month suspensions of mortgage payment plans for borrowers facing severe financial hardships. Some of the challenges for governments include securing the flow of credit for borrowers and lenders, maintaining liquidity in the mortgage markets, and protecting the system against future catastrophic events. To this end, entities are adjusting the terms of their underwriting policies in order to manage risk appropriately, as well as using advanced analytics to calculate risks and predict hardships, including liquidity and ongoing swings in borrowers' credit health. Also, a multi-stakeholder response is needed to counter the adverse effects of the pandemic, creating the macroeconomic conditions of recovery as well as internal growth. In the region, four cases are of special interest regarding the financialization of housing: Chile and Mexico as the forerunners and producers of massive housing programs, while Brazil and Colombia profited from conjunctural moments in the history of their countries.

Chile and progressive housing programs

In the early years of Augusto Pinochet's regime, the Ministry of Urban Development and Housing (Ministerio de Desarrollo Urbano y Vivienda, MIDUVI) heralded a new housing model under neoliberal policies that aimed to remake the conditions for the operation of urban land markets, claiming that "the State will endorse and engage in the creation of an open market in housing [while] the responsibility for production will be taken on by the private sector" (MINVU, 1979, p. 7). Later, as a countercyclical approach to the economic crisis of 1982, MIDUVI created a system that interlinked public subsidies, savings, and credit, enabling construction companies to build housing developments across the country. Despite the discussion of the neoclassical economic angles on scarcity, the markets' rationale would solve those contradictions:

> For an extended period of time it has been insisted upon that urban land is a scarce and irreplaceable resource, which has contributed to its price suffering frequent distortion in the marketplace by artificially restricting the supply. Present policy, on the contrary, is based on the principle that land is not a scarce resource, but that *its apparent scarcity is provoked in the majority of cases by the inefficiency and rigidity of the rules and legal procedures* applied up until now in order to regulate the growth of cities.
>
> (MINVU, 1979, pp. 18–19)

Between 1990 and 2005, the Chilean government directly subsidized the purchase of 1.2 million new housing units – roughly a quarter of the new dwellings built during this period – benefiting mainly families at the lower end of the income scale (UN-Habitat, 2009, p. 1). This intervention was part of a strategy to reduce the accumulated deficit of social housing units by boosting demand and guaranteeing a steady supply, resulting in the reduction of the housing deficit from 23% in 1992 to 10% by 2011 (Fig. 4.2). Even if the social housing mechanisms put in place in Chile have been praised and inspired similar policies in the region, it has faced a number of criticisms. The quality of the housing stock has been questioned since they have not allowed expansion, improvements, or adaptations to the changing needs of the households. To this end, Rodríguez and Sugranyes (2012, p. 57) contend that at the turn of the twenty-first century in Chile "per capita income had doubled while inequalities grew deeper and social networks disappeared … but the production model as well as the housing typology prevailed, more for the worse than for the better." Moreover, homeowners have not been able to take advantage

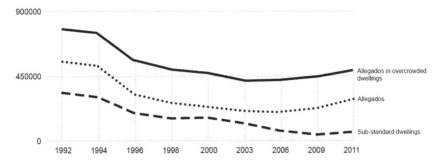

Fig. 4.2 Evolution of the housing deficit in Chile by component, 1990–2011 (households).

Source: Ministerio de Desarrollo Social, 2013.

of it as an investment: prices have stagnated while those of middle and high-end properties grew 14 times between 1990 and 2004 (Rodríguez & Sugranyes, 2012, p. 58). Therefore, even when access to decent housing conditions improved the living situation of households, their investment did not produce gains over time, or result in social mobility, since wages remained unchanged while debts increased.

The commodification of social housing in Latin America brings together the State, the market, and society through a financial rationale, aimed at public policies that enhance governance, urban competitiveness, and individual approaches to the housing question. During the early 1990s, the World Bank introduced the idea that the role of the State was to enable the operation of the housing market through the *smart* use of subsidies, effective cost-recovery strategies, and the provision of mortgages and guarantees that allowed flows of capital from the private sector to the real estate markets. To this end, the State is in charge of protecting the financial system's integrity through the allocation of subsidies, securing savings and institutional funds, as well as advancing regulations to enable the operation of capital markets.

As a result of the preeminence of the Washington Consensus, the previous rationale revolving around rights and equity in Latin American countries was reframed to individualizing social demands, enabling the operation of housing markets but also favoring the privatization of services such as water, electricity, and garbage collection, as well as social services, framing rights as inefficient and directing welfare mechanisms only to the most vulnerable groups. From 1993 on, the World Bank set the social housing agenda, advocating for its capabilities to become an important sector of the economy by endorsing markets to solve housing needs, mobilizing private capital, and restricting public subsidies (World Bank, 1993).

In the case of Chile, a system of vouchers for lower-income citizens to purchase dwellings was put in place, thus creating a low-income housing market that was fueled by public money along with flexible urban planning controls, incentives, and tax exemptions, as well as expanding land banks around cities where the markets would set the prices for the best locations. According to Rodríguez and Sugranyes (2005, p. 15), during the early years of the democracy in Chile, the new left – which was to rule the country for the next two decades – was more concerned with the construction of large social housing developments, which, beyond their quality, location, or environmental provisions of the settlements, were important because they scored high in public approval. To this end, urban planning constraints were lifted and deprived of any comprehensive approach, defining urban expansion through its profit-earning capacities (Rodriguez & Icaza, 1993, p. 3), hence creating dispersed and disconnected territories that were immersed in wide-ranging processes of diseconomic proportions.

Housing finance in Chile has seen a consistent development trend in the past three decades, enhanced by the relative economic stability experienced in that period, sharp financial supervision of financial markets, pension reforms that allowed the use of institutional funds, and the creation of an inflation-indexed unit of account. Housing finance operated through mortgage bonds (*letras de cambio*) until 2004, and from then on, mortgage credits backed by mutual funds started to take hold of the market. At the same time, the national public bank of Chile (BancoEstado) started to finance social housing mortgages with credit risks that were considered as subprime for commercial banks.

During the first decade of this century, Chile's public expenditures on social housing as a percentage of GDP were one of the highest in the region, accounting for an average of 100,000 housing subsidies per year, halving the historic deficit of social housing, and yet increasing the *qualitative* deficit associated with space overcrowding, while limiting the possibilities of upgrading and expanding the units. What has been distinctive of the Chilean experience has been the use of two major conduits to finance social housing: the first are mortgage bonds, which were issued to lenders to obtain more resources to fund mortgages, while financial institutions sold them on the stock exchange (Bolsa de Comercio de Santiago); the second conduit was the endorsable mortgage mutuals, composed of liquid credits that are endorsed to creditors and administered by banking institutions.

The US subprime crisis had a direct affect in Chile, resulting in an economic recession in 2009, which prompted credit restrictions and the escalation of mortgage portfolio risks that raised default rates, even if concentrated in the national bank, BancoEstado. After the crisis, *leasing* was proposed as an alternative mechanism to access social housing, where debtors signed a contract with the financial institution that included a buying option, but it did not turn out well since interest rates remained significantly higher than those applied to mortgages. As in other countries in the region, default rates went back to normal after a few years, and in order to improve its delinquency rates, BancoEstado started using cross-references to assess risks with historic information on clients such as personal and fiscal debts, reducing delinquencies to 200 cases out of a pool of 500,000 housing loans.

According to Razmilic (2010), 60% of the housing stock in Chile was acquired through direct housing subsidies in the last 30 years, representing 3.4% of the annual national fiscal budget for that period. The Ministerio de Vivienda y Urbanismo de Chile (MINVU) had been in charge of social housing since the late 1970s, intervening in the production of housing as well as providing credit for mortgages and subsidies for the poor. However, since the early 2000s, the Ministry favored subsidies, allegedly using efficiency as its justification, creating the Fondo Solidario para la Vivienda (FSV) as a fund to subsidize housing for the most vulnerable population through vouchers for the total value of a unit (FSV-I), while a second subsidy bond was directed to a middle-income population, covering the difference between the sale price and the mortgage plan (FSV-II).

In the last decade, Chile allocated 0.52% of its GDP to homeownership provision, which is considered to be high even for OECD countries' standards, offering first-time buyer grants, rent-to-buy schemes (leasing), as well as other subsidies for housing maintenance, home upgrading, and tax-deduction schemes (Fig. 4.3). A major leverage for social housing finance has been the intervention of institutional investors in real estate markets. Another interesting undertaking was the introduction of Social Real Estate Management Entities/Entidades de Gestión Inmobiliaria Social (EGISs), which were created to oversee the process of bringing together pools of applicants, housing projects that could benefit from the DS174-I subsidy provision, and secure land banks for the construction of the projects. The entities could be public or private (for profit or not), while the delivery of the subsidized dwellings was under the authority of an autonomous entity called Housing and Urbanization Services (Servicios de Vivienda y Urbanización, SERVIU), responsible for the technical inspection of the works and paying fees to EGISs for their services in accordance with regulatory values. Private EGISs were not-for-profit entities associated with private foundations, international NGOs, and religious organizations, while the for-profit version operated within an entrepreneurial rationale that created value out of their managing capacities to carry out the complete housing production process. Public EGISs are associated with municipalities and follow a subordinated logic, specializing in less attractive undertakings such as housing projects in rural or small urban areas.

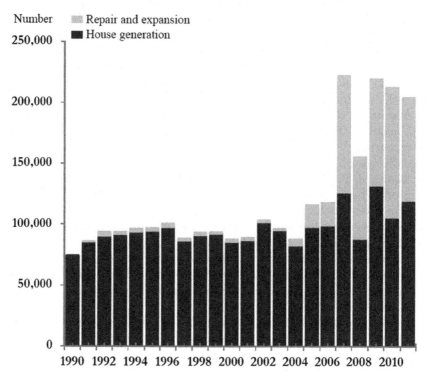

Fig. 4.3 Type of grant awarded.
Source: Ministerio de Vivienda y Urbanismo de Chile, MINVU, 2009.

A complementary program called DS49 targeted low-income households in conditions of social vulnerability and social deprivation, aimed at granting access to homeownership either through construction of, or purchasing, a home. The total cost of the unit could not be higher than USD 33,600, while recipients had to make a one-off contribution of USD 420 to complement the subsidy. Since this was not a mortgage scheme but a subsidy, households were selected through a multidimensional vulnerability score that measured needs and risks, requiring recipients to move into the dwelling and reside in the municipality where it was built, so as to prevent the uprooting of households from their original communities. Additionally, subsidies for densification projects aimed to target better locations and land uses for the subsidized dwellings, even if the real estate markets rationale favored higher profits. The innovation in this program was that housing projects had to improve the environmental quality of subsidized housing through the provision of public spaces, even if that increased the production costs and reduced profit margins.

In the last two decades of the twentieth century, Chile built the largest volume of low-income housing units in its history, accounting for more than one million housing subsidies, where three-quarters of households were built in deprived sectors where social, educational, and economic disadvantages prevailed (MINVU, 2014). However, this policy led to rapid decline of the units as well as furthering environmental degradation and social decay, and by the mid-2000s upgrading initiatives had to be implemented. Among them, new urban and housing policies addressing quality upgrading and social integration were introduced that were aimed at articulating "three levels of intervention: housing, neighborhood and the city, in order to assemble both sectorial processes and investments" (MINVU, 2009, p. 18), granting with it a territorial perspective that looked beyond the housing question alone.

Along these lines, the Neighborhood Recovery Program DS14 (MINVU, 2016) targeted the retrofitting of public spaces and local environments using the neighborhood as a scale of intervention. Similarly, the Regeneration of Housing Developments Program DS18, aimed at "building or improving green areas, amenities and roads, construct[ing] or upgrad[ing] social dwellings, formalizing social organizations and co-ownership schemes" (MINVU, 2018, p. 44). Interestingly, these new approaches involved particular strategies of intervention that included urban regeneration plans and projects, taking a more comprehensive stand and acknowledging the intersectional and multidimensional nature of urban and environmental problems.

After the major earthquake and tsunami of 2010 in Chile, real estate markets developed swiftly even when the cost of building materials escalated during the reconstruction process. The booming economy was led by the copper industry, which was credited for investing its surpluses in real estate, land, and even rental markets across the country, particularly in the capital city. However, the conjunction of these events had a negative impact on the production of social housing since the possibility of improving the architectural solutions and location of the units vanished as land prices absorbed the circulating capital.

Originally, the System of Housing Subsidies DS40 required prospective applicants to file a group application to secure a voucher for a housing project, but from 2011 an amendment allowed individuals to apply for permission to purchase a dwelling directly from the developer.[1] In addition, a middle-income subsidy was introduced to capture a sector of the population with the economic capacity to qualify for a mortgage loan. It is important to note that these schemes created dysfunctional outcomes, leaving 30,000 voucher recipients waiting to find social housing units in Santiago in the early 2010s (Imilan Ojeda, 2016), while another 30,000 families lived in informal housing (*campamentos*) across the country during that same period (Fundación Techo, 2013). As we have seen, wider-scope programs have been put in place to address environmental and social problems in areas concentrating disadvantages, vulnerabilities, and decay, where the articulation of multilevel investments was not fully accomplished and interventions to address social ailments such as organized crime were left incomplete, and yet the introduction of a territorial interdisciplinary and citizen-based approach was unprecedented in Chilean urban policy.

Mexico and private housing corporations

Mexico went through important constitutional changes during the 1990s (Art. 27 in 1992), enabling the transformation of formerly socially owned land (known as communal and *ejidal*) into privately owned property. This opened up land markets to the construction of housing developments, producing 5.8 million units in that decade and 10 million in the following (INFONAVIT, 2014, p. 16). Using financial mechanisms which we will discuss later in this section, over the last decade INFONAVIT issued half a million mortgages a year and yet the rate of vacant housing went up 14%, accounting for 6.1 million abandoned units across the country, as well as another 2.5 million temporal-use second homes (INEGI, 2020). This case of vacant stock of social housing developments is crucial for understanding the impact of real estate markets in the production of space because of its economic, social, and sustainable implications on the territory. First, vacancy rates imply that mortgages were defaulted, leaving financial institutions, investors, and housing developers negatively affected by the failure of these major investments. This also had deep economic and social consequences for the low-income population, whether their properties were foreclosed or remained within a declining environment where the value of properties plummeted, the housing quality decayed, and insecurity rose, in the process reducing social interaction and the environmental quality of their housing developments. This situation hinders even further local governments' capacity to deal with the provision of infrastructure, transport, facilities, and basic services to these developments, since they are dispersed along the peripheries. Therefore, local authorities have to deal with these shortages in the short and long term (facing the paradox of having a massive stock of empty units), encountering an accumulative process of decay and devaluation of large sectors of the built environment that sooner or later will have to be addressed.

After two decades of neoliberal low-income State-led housing practices, INFONAVIT conducted an assessment of the vacant housing expansion to determine the possible causes of this situation. Location represented a major inconvenience since developments were far from jobs and services, creating economic instability that constrained the amortization of mortgages. Also, an important percentage of borrowers took the decision to engage in a mortgage scheme to invest in a second home, an asset that could later be sold or rented. Second, the process for

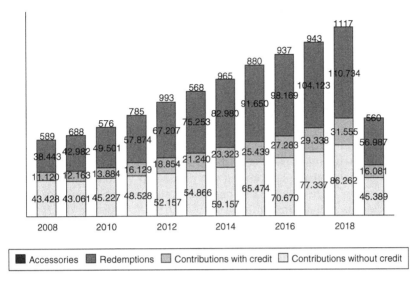

Fig. 4.4 INFONAVIT contributions and loan repayments (billions of pesos).
Source: INFONAVIT, 2019.

underwriting a mortgage did not comply with sufficient information on the contract specificities, its location in relation to jobs and amenities, the construction standards, or public security, while the prospect of using their savings, which otherwise would not be possible to access, propelled a general trend of securing mortgages for the low- and middle-income population (Fig. 4.4).

Let us now turn to how financialization introduced new patterns of capital accumulation in Mexico, which entailed approaches and rationalities that diverted from earlier socioeconomic challenges, centering instead in technical procedures and operative formulas. We have been suggesting that financial markets are producing rapid transformations in the built environment that generate social and cultural transformations and increase spatial fragmentation, furthering social segregation and economic polarization. The financial rationale is presented as a mechanism to transform commodities, property, debt, and expectations into tradable assets, stocks, and certificates. The power behind this configuration relies on the promise of large returns in the short run, fueled by self-fulfilling prophecies of growth, despite the possibility that this may later lead to collapsing the whole socioeconomic structure.

Past economic crises confirm their cyclical nature, where boom and bust episodes precede a new configuration that leads to the next meltdown. In this sense, Mexico has experienced a number of economic crises in the last four decades that have shaped the sociopolitical framework in which the country stands today. These crises may be traced back to 1982 when the national currency lost 470% of its value, after which the World Bank and the International Monetary Fund offered to bail out the country in exchange for harsh structural reforms of the type inscribed in the Washington Consensus (Soto, 2013, p. 62). In the next financial crisis, in 1994, investors transferred their capital abroad while dismantling the economic infrastructure, which led the way to further privatization of formerly State-owned enterprises, public services, and infrastructure, as well as denationalizing the banking system. This scenario framed financialization for an extended period of time, as economic and operational restructuring took place in order to bring about the promised changes. Social housing epitomized the process in which formerly State-led institutions were reshaped to perform according to a financial rationale beyond its social constraints.

In 1963, near the end of the period known as the "Mexican miracle" of stable economic development (1940–1970), the State created the Housing Operation and Banking Finance Fund (Fondo de Operación y Financiamiento Bancario a la Vivienda, FOVI) to provide low-interest mortgages for workers and state employees. A decade later two institutes were founded to expand the constitutional right to decent and affordable housing (INFONAVIT and FOVISSSTE) as long-term saving schemes, while undertaking the production process of the housing units and establishing standards in the areas of construction, planning, and design. During the economic crisis in the early 1980s the institutes reformulated their operations, indexing their interest rates to inflation and applying stricter loan recovery policies as demanded by the World Bank (Monkkonen, 2011, p. 675). Along with the structural reforms demanded by the neoliberal economic framework, the institutes transformed their role in social housing provision to promote financial efficiency, enabling the production of housing by subsidizing the demand as well as assuming the role of financial intermediaries through granting mortgages to lenders in a market dominated by major housing developers such as URBI, GEO, HOMEX, ARA, SARE, and Consorcio HOGAR (Valenzuela Aguilera, 2017).

The privatization of the banking system and capital flows in the 1990s led to the rapid growth of mortgage loans but a sharp increase in interest rates of up to 74% following the 1994–1995 financial crisis, and led to extensive default rates that nearly took the banking system to collapse. After this crisis, and with the privatization of communal and social land, new schemes of housing production based on the volume of units rather than the creation of sustainable urban environments for the population of lower and middle incomes were introduced. The new rationale trimmed former social provisions including subsidies, while eliminating mortgage and title insurances and instead relying on the MBS framework. As the major national banks had been sold to international financial institutions,

FOVI was reframed as a mortgage *securitizer*, while the institutes co-financed the loans with funds from the World Bank, FOVI, and second-tier development banks such as BANOBRAS (Puebla, 2002, p. 77). With the banking system restructured, FOVI was transformed into the Sociedad Hipotecaria Federal (SHF) to endorse the securitization of mortgages through new vehicles called SOFOLES (Sociedad Financiera de Objeto Limitado/special-purpose vehicles) and SOFOMES (Sociedad Financiera de Objeto Múltiple/multiple-purpose vehicles), which engaged lenders outside the formally registered jobs requirements by the institutes (Fig. 4.5). Instead, SOFOLES issued RMBSs to finance their loans that were backed with SHF funding, allowing lenders to combine institutional resources as down-payments to issue market-based private mortgages, a combination resulting in nearly 850,000 mortgages between 2000 and 2013 (Marosi, 2017).[2]

Since the turn of the twenty-first century, the Mexican government has tried to establish a framework to enhance the primary and secondary mortgage markets, stressing the importance of financial mechanisms such as improving contract enforcement, streamlining foreclosures as well as property and creditor information, private hedging, improving registers, and adopting market-oriented mechanisms as a rationale to justify their efforts (Fig. 4.6). Deregulation allowed the creation of more than 120 SOFOLES by 2006 and some of them became more prominent than their financial bank counterparts, while being funded by the International Financial Corporation, a subsidiary of the World Bank Group. However, following the US subprime crisis the next year, they were faced with a 53.5% default rate; State funding ended for such institutions, which later led them to bankruptcy.

Originally, when FOVI was founded by the central bank in Mexico (Banxico) along with the World Bank, it was as a second-tier banking institution, providing funding and guarantees (up to 45% of loss-given default) to banks extending mortgages to targeted households and low-cost housing developers (IMF, 2008). Following the US scheme of government-sponsored agencies created to buy mortgages from loan originators, the Sociedad Hipotecaria Federal (SHF) substituted FOVI and established certain standards to pool mortgages and back their securitization in the secondary mortgage market. Just as Fannie Mae and Ginnie Mae provided partial credit mortgage guarantees acting as a second-tier bank for private entities, SOFOLES' role was to support financial institutions issuing RMBSs through the capital markets by guaranteeing the timely payment of RMBS obligations (Fig. 4.7).

Credit enhancements by the SHF were known as BORHIS (*bono respaldado por hipotecas*/mortgage-backed securities). They were bought in the primary and secondary markets at a price based on particular prepayment and default assumptions (taking up to 35% of losses), positioning SOFOLES as the issuer of 60% of total outstanding RMBS stocks (followed

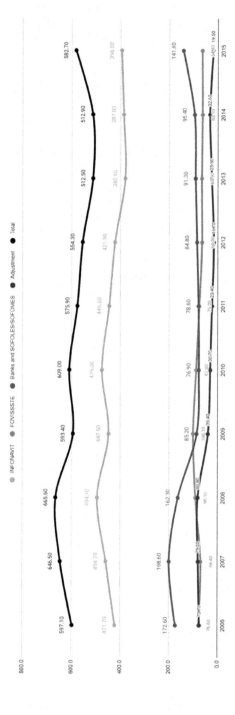

Fig. 4.5 Number of mortgage loans to households (in thousands).
Source: BBVA Research, 2016.

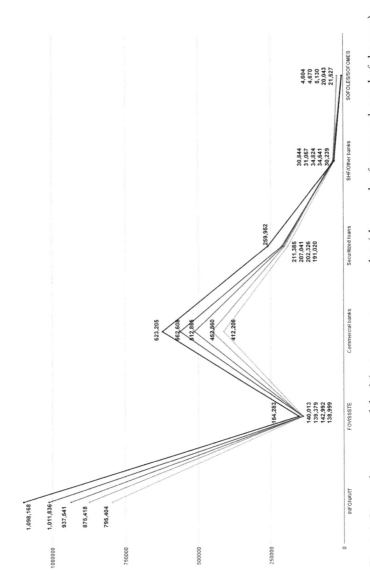

Fig. 4.6 Size and structure of the Mexican mortgage market (thousands of pesos at the end of the year).
Source: Banxico, 2011–2016.

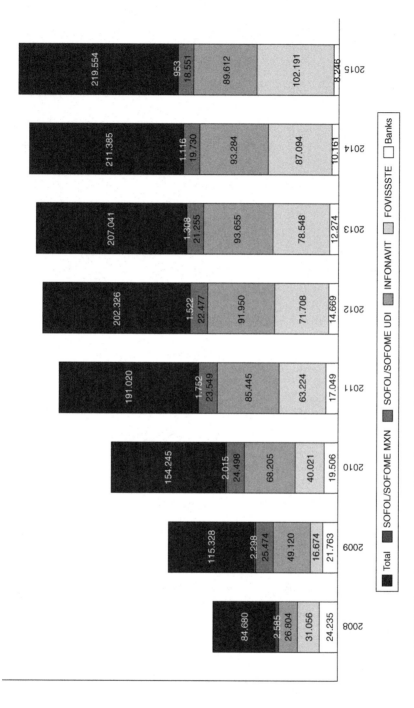

Fig. 4.7 RMBS market structure and development (millions of pesos).
Source: Moody's, 2016.

by INFONAVIT with 27%), while institutional investors such as domestic pension funds and insurance companies dominated the structured assets in the financial markets. Credit enhancement through financial guarantees permitted higher investment ratings, which otherwise would have been difficult to obtain, as well as higher yields than comparable government securities, therefore attracting international investments in RMBSs.

It is important to understand that the subprime crisis in the US did not bring down the housing system in Mexico at the time. It was the national infill development policies that were put in place that disrupted the previous expansive development model and accelerated its collapse. The incoming administration of Enrique Peña Nieto (2012–2018) directed subsidies to housing developments in more central locations, meaning that many previous land reserves were rendered useless for further urbanization, contributing to the decline and bankruptcy of some of the major construction companies in Latin America. Even when the subprime crisis affected a developer's performance in the stock market, mortgages in Mexico kept growing in volume while also increasing their contracted amounts, targeting middle-income borrowers. Peña's administration launched a national program for urban development and housing to establish higher densities through infill development strategies – locations already served by infrastructure and services, where there were job opportunities. This was done primarily through the provision of tax exemptions and fiscal incentives.

Since that last crisis, the SHF has limited its operations in the second mortgage market and oriented its activities to hedge risks instead of enhancing the second mortgage market, while institutes shifted their credit indexation from minimum wages or *unidades de medida y actualización* (units of measurement and upgrading) to Mexican pesos. Also, financial products diversified, addressing ecological features, renovation, improvements, as well as rental investments, yet leaving behind the self-construction practices that prevail in the informal market: independent workers, family enterprises, local merchants, etc. (Fig. 4.8). Instead, housing-finance institutions concentrated on their affiliates, allowing them to provide generous conditions for the RMBSs they issued to cross-subsidize households that earned less than the equivalent of 6.5 times the minimum wage, a practice that has been questioned as a market-inefficient provision.

Before the subprime crisis, the SHF funded and allocated credit enhancements to SOFOLES/SOFOMES in order to foster the secondary mortgage market. However, later on investors lost confidence in the quality of their portfolios, which made the sector disappear after the crisis. Then these entities were converted into banks under new regulations and requirements such as credit provisioning, capital adequacy, risk diversification, and accessible information. As we have suggested, the housing crisis unraveled in Mexico with the change in urban policies that no

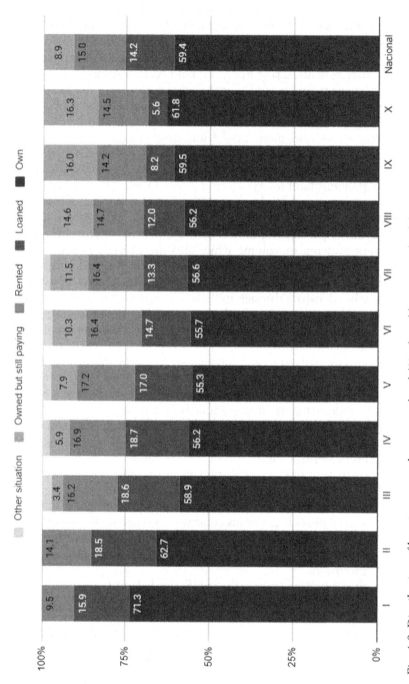

Fig. 4.8 Distribution of housing tenure by income level (% of total by income decile).
Source: ENIGH, 2018.

longer favored peripheral developments, which directly affected some of the biggest housing developers that had already purchased large areas of land. These land reserves were rendered inadequate since the new tax provisions and regulations would not allow housing developments in such areas, so, as noted above, the corresponding decline in value led the companies to bankruptcy.After the housing crisis in Mexico, more prudential lending rules were established requiring minimum underwriting criteria, including verified income and credit history as well as a recent appraisal. The rules also put housing institutes in a privileged position for repayment mechanisms since recovery is secured through direct payroll deduction. Nonetheless, non-performing loans have still increased significantly in different periods (for instance, in relation to the vacant housing stock) (Fig. 4.9). According to an INEGI survey (INEGI, 2014), out of 49.5 million working people, 29 million (59%) were not affiliated with a social security system, while 18 million (36%) were affiliated to IMSS/INFONAVIT and 2.3 million (5%) to ISSSTE/FOVISSSTE.

New provisions for financial markets have been in operation since 2016 with the Basel III framework setting standards requirements for risk management functions such as internal rating of loans, mortgage risk assessment, definition of delinquencies, and methodologies of provisioning aligned with banking rules. This framework seeks to ensure that financial institutions will have stable resources to find assets even in the case of a market disruption for a period of at least a year. To this end, the Basel Committee on Banking Supervision (BCBS) is looking to promote the concept of simple, transparent, comparable securitization (STC) that allows investors to evaluate risk and return. The simplicity entails the homogeneity of the underlying assets and clear deal structures, where information has to be disclosed, enhancing confidence among investors, establishing quality standards for the configuration of securities, allowing no more than three tranches, simple waterfall, limited credit events, direct allocation of flows, certification systems, public registering, and a sanctions regime for non-compliance (Basel Committee on Banking Supervision, 2016).These reforms are expected to have a positive impact on the real estate market, lowering interest margins and rates as well as reducing the fees and expenses charged by originators. Sociedad Hipotecaria Federal as an RMBS market maker endorsed the creation of the HITO technological platform specialized in mortgage portfolio management, an initiative to promote private RMBSs (especially by SOFOLES and SOFOMES), acting as a multi-users conduit and compliance monitor (Fig. 4.10). When the diversification of mortgage products took place, it increased home improvement loans by 40% (although only representing 5% in local currency value), reverse mortgages for senior citizens, rental housing, and the percentage of microcredits to purchase low-income housing. Concurrently, urban information platforms have been developed

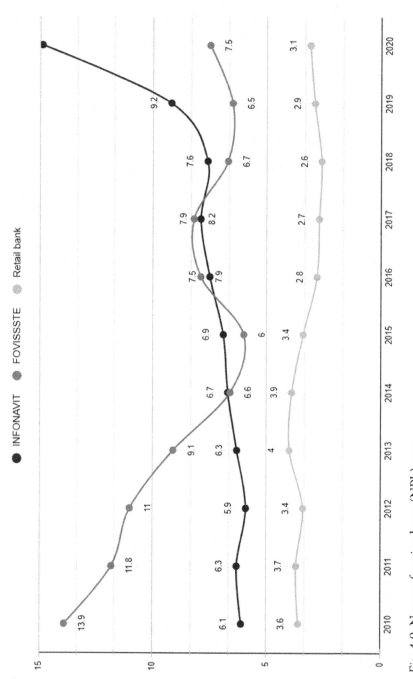

Fig. 4.9 Non-performing loans (NPL) rates.
Source: AMB, 2020.

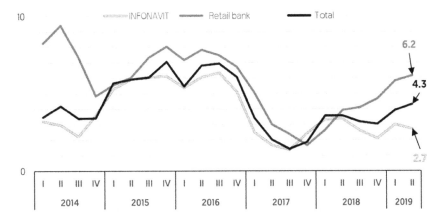

Fig. 4.10 Growth rate of the mortgage loan portfolio (annual percentage change). Source: INFONAVIT, 2019.

and will be useful for financial institutions to assess the potential of new developments, property values, and possible housing bubbles through a National Registry of Territorial Reserves (RENARET), a national information and housing indicators system (SNIIV) to monitor and assess the housing markets, as well as a Housing Registration System (RUV).

The rising mortgage payment default curve prior to the pandemic also indicates that the mortgage system had encountered structural problems in which the capacity of INFONAVIT to secure performing loans had diminished in recent years, calling for a new system of mortgage credits. However, social housing has been particularly sensitive to economic upheavals and with the COVID-19 pandemic in 2020, mortgage delinquency rates increased 14.93% in INFONAVIT, 3.13% in the general banking system, 2.60% with BBVA, and 1.28% at BANORTE, which reveals a complicated scenario for the upcoming years, as can be seen in the canceled credit enrollments at INFONAVIT shown in Fig. 4.11.

Brazil and *Minha Casa, Minha Vida* program

The financialization framework has been instrumental for allocating global capital flows and funding spatial fixes in cities (Harvey, 2003). This rationale suggests that developers should rely on capital markets rather than traditional banks and institutional funds, and yet it is not clear if these procedures have served as a driver for social and economic change (Pereira, 2017). In the case of Brazil, this process targeted large-scale, subsidized social housing mechanisms along with the reconfiguration of the real estate financial sector regulations, as new funding schemes were

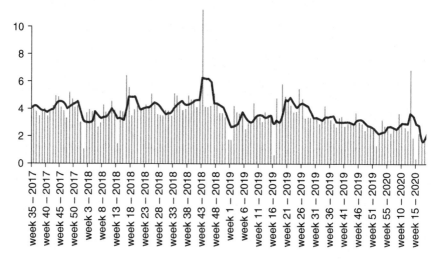

Fig. 4.11 Canceled credit enrollments.
Source: INFONAVIT, 2020.

introduced in the last decades. Even when some authors suggest that the country did not have a Fordist welfare state to dismantle with the emergence of neoliberal policies in the first place (Pereira, 2017, p. 605), the role of the government in enabling the creation and recreation of markets through financial instruments was crucial (Fig. 4.12).

As with the rest of the Latin American region, Brazil was subject to the structural adjustment agenda imposed by the World Bank and the International Monetary Fund, which entailed privatization mechanisms in which market actors were encouraged to prioritize profit maximization. This enabled the configuration of a new real estate financial system, which established a regulatory framework allowing the issuing of financial instruments such as the certificates of real estate revenues (CRIs) that were central to the securitization process prior to the subprime crisis (Royer, 2009). These certificates were structured like US mortgage-backed securities, allowing the financialization of real estate beyond traditional sources of funding through the Housing Financial System (Sistema Financeiro de Habitação, SFH). The system was structured through two different funds, the Fundo de Garantia de Tempo de Serviço (FGTS), which is the mandatory saving scheme where workers deposit 8% of their income, and the Sistema Brasileiro de Poupança e Empréstimo (SBPE), a savings and loan system created to fund real estate mortgages with better conditions than RMBSs while also offering low-interest financing (Cardoso, 2013). Brazil was a latecomer in the securitization of mortgages, real estate, and the consumer debt boom, and yet this scheme was put in place for the

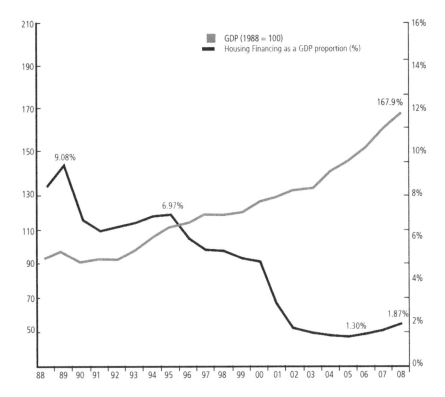

Fig. 4.12 The Brazilian Housing Financing System as a proportion of GDP.
Source: BACEN, 2022.

long run. For instance, between 2006 and 2007, 25 construction companies went public on the São Paulo Stock Exchange (BOVESPA) and the
following year these companies were worth more than USD 20 billion
(IBGE, 2006).

While FGTS and SBPE offered cheaper credit schemes during the
second half of the 2000s, regulatory adjustments and tax exemptions
made the CRIs more compelling for investors in that period. Also, a similar instrument to a leasing contract protected creditors, holding them as
legal owners of the property until the entire loan was paid, in the process
allowing foreclosures to take place. In that decade, major construction
companies in Brazil profited from the liquidity surpluses of global capital markets, the growth of domestic GDP, and the declining interest
rates that peaked in 2007 (Sanfelici, 2013). Several of these companies were listed on the stock exchange through Initial Public Offerings
(IPOs), enabling easier associations with private equity and pension
funds, which later became prominent in their management bodies and

had a greater say in their business strategies and corporate governance. In the middle of these structural shifts, real estate companies adopted a more corporatist rationale, which was shareholder-oriented, looking to enhance the stock market value through the acquisition of large land reserves to build housing developments. Also, a major growth engine was greater access to mortgages for the middle- and lower-income population, which allowed companies to build beyond the housing demand and disregard debtors' solvency to pay their mortgages. Financial capital inflow leveraged real estate markets during the housing bubble, yet they also proved critical when downturns took place and altered housing development dynamics.

In order to face the upcoming economic crisis of the Great Recession the Brazilian government launched the *Minha Casa, Minha Vida* (MCMV) program in 2009 to enhance housing demand while also serving as a major incentive for the construction sector. The program introduced housing as a commodity into the circuits of capital accumulation in 2009. Between 2003 and 2014 the MCMV raised the total value contracted by the FGTS from USD 1 billion to USD 28 billion, while the SBPE increased from USD 0.7 to USD 47 billion, and another USD 14 billion was allocated from the general budget of the union to finance the program (Cardoso et al., 2017). The program targeted three income groups using significant subsidies, which covered nearly 83% of the total cost per unit for the lowest income group (monthly wages up to USD 800). The middle-income group (up to USD 1,600) received an initial grant worth 17.5% of the total cost, as well as public insurance, while a third group (up to USD 2,700), obtained the insurance but not the initial grant. It is important to point out that the housing deficit in Brazil was estimated by the João Pinheiro Foundation (2021) to be 6 million households in 2019, 74% of which were concentrated in the lowest income group, which were targeted by the MCMV program, and yet the program built 62% of the housing units for middle-income groups (Furtado et al., 2013).

The program also enabled the transformation of informal households into formal homeowners, although this often required relocating them to peripheral milieus, weakening the social fabric and communitarian structure that may have been in place, and replacing it with a new social *substratum* now dependent on the authorities, as mortgages remained mostly under the control of public institutions. The first projects were developed by private companies and underwritten by the Caixa Econômica Federal (CAIXA), which protected them against default risk, where loans were granted at submarket interest rates, and owners benefited from the public insurance system (FGHAB).

During former president Lula da Silva's administration, unemployment rates were cut in half and informality reduced by 3 percentage points, while the GDP doubled during the first two decades of the twenty-first century, elevating 50 million households to the middle-income classes.[3] Also, investments in welfare policies regarding education, health, sanitation, and housing reached a record of USD 292 billion in 2009, accounting for 15.8% of GDP. At the time, Brazil counted 57.6 million households, where three-quarters were homeowners and 86% lived in detached homes, although only 55.4% were connected to urban sewer systems. This, along with a historic deficit of 6 million dwellings (equivalent to 10% of the housing stock), supported the potential for a privately financed social and middle-income housing market (UN-Habitat, 2013, p. 33), for which the MCMV program targeted low-income groups, aiming to contract one million units by the end of 2010.

Yet, according to the Demographic Census 2010,[4] there were already at least 6 million vacant housing units, 70% of which were in urban areas. The number of dwellings was equivalent to the housing deficit referred to above, although their legal occupancy status was uncertain. This replicates the situation in other countries in the region such as Mexico, where 6.1 million vacant housing units were reported in the 2020 census, while the historic housing deficit was estimated to be around 8.9 million dwellings. This recurrent phenomenon emerges from a mismatch between the housing supply and the income level, where at least 70% of the urban population could not afford to access housing within the formal housing market, calling for housing subsidies schemes for low-income groups worth USD 18.4 billion, with a strong redistributive component as part of the crisis-response policies. At this time, the MCMV had a non-refundable budget containing 75% of the total investments in subsidies from the federal budget of the union (Orçamento Geral da União) as part of upfront subsidies targeting lowest-income households and allocated through the Residential Leasing Fund and the Social Development Fund.[5] In the case of refundable funds that were originated from the workers' Severance Fund, financial operations required return payments through installments that were calculated as a ratio of the household's income (10% over a decade), being more of a sign of commitment than a real cost-recovery mechanism. The MCMV program required as a condition for a CAIXA mortgage contract that basic infrastructure provisions in the housing developments had to be included. However, it was the municipality's responsibility to extend and provide public facilities such as education, health, and leisure, as well as basic services such as electricity, roads, sanitation, and drainage systems, public provisions that should not be taken for granted.

As noted above, it has been written that the production of urban space reflects an underlying configuration that responds to particular logics and interests of the agents involved: the State, the real estate market, and the population (Castells, 1974; Lipietz, 1974; Topalov, 1974). However, in the last decades finance has risen as a parallel rationale that orients the investments in the urban realm through the maximization of profits. Even when this configuration does not include the informal production of space (self-help housing, but also associated amenities), it accounts for a major part of the formal housing stock in Latin American cities, changing with it the procedures as well as the use of public resources to fund large-scale development projects.

In the case of Brazil, direct access to financial markets was granted by the CAIXA or Federal Savings Bank through its *cartas de crédito* (letters of credit) program, which allowed access to public funding for housing without the mediation of developers or public housing agencies. As a result, in the MCMV program construction companies served as brokers, connecting landowners, financial institutions, and consumers. And in other ways, as we have suggested, the State played a central role in enabling mechanisms that enhanced the financialization of social housing production. It exempted financial gains from the securitization of real estate assets from income tax and enforced default contingencies, as well as establishing the National Housing Finance System, which provided low-interest rates for loans to finance mortgages (Royer, 2014). Around 2014, the CAIXA created a special entity to serve construction companies and developers called the Corporate Bureau, which helped increase the volume of available resources for housing finance, providing guarantees to investors in that sector. These provisions attracted international and domestic capital for housing production along with the *Sistema Financeiro de Habitação*, blending the two financial sources.

The Housing Financial System (SFH) was created in the early 1960s as a real estate credit entity and was later complemented by a parallel system that could operate with a private rationale called the Real Estate Financial System (SFI), operating along financial and capital markets parameters, and notwithstanding weak investment vehicles could still render the secondary markets ineffective. The SFH uses private resources administered by the State through the CAIXA, which is the main commercial bank responsible for issuing 73% of housing credits in Brazil, while the Central Bank manages the treasury funds and oversees the whole financial system, and the Securities Commission controls and regulates the operation of capital markets. The SFH holds banking private accounts (*cadernetas de poupança*) from which funds are deposited monthly into a *Fundo de Garantia por Tempo de Serviço* (FGTS), which is a governmental retirement fund exempted from income tax; the second instrument are the *letras de crédito imobiliário* (LCI) which are savings funds used as the principal source for funding debt securities through qualified investors that have a significant portfolio in the market.

The SFI is funded through certificates of real estate receivables (*certificados de recebíveis imobiliarios*, CRI) as investment instruments that securitize mortgages in the secondary markets, capturing the flow of deposits in pension funds to finance the acquisition of homes. However, the certificates used as instruments for mortgage securitization remain in the investors' portfolios pending maturity and not as tradable securities, hence restricting their placement in the financial markets. As a result, mortgages are internalized in banks' portfolios, which means that the financial housing system obtains no leverage, nor produces liquid securitization instruments that could be attractive for secondary financial markets. Under this scheme, the FGTS holds savings accounts where the employer deposits 8% of the workers' wages every month, which are only unlocked for the acquisition of housing or in the case of retirement. Since those funds are administered by the CAIXA, this institution can act as a second-tier bank for the provision of resources to purchase mortgages originated in a first-tier bank specialized in housing finance. Therefore, a major difference with other financial systems in the Latin American region is that the FGTS were not leveraged, since the returns of the mortgage payments go to restore funding to run the system and are not subject to securitization.[6]

In the early 2000s, major construction companies in Brazil went public, which had a significant impact in the real estate market, expanding their operations financially and geographically across the country. The MCMV program prompted private-sector investment in the provision of social housing, as firms sought to benefit from tax exemptions directed to companies catering to low-income families, which in the process was attracting large developers due to the low impact on their cash flows or risk exposure, while profiting from economies of scale. Nonetheless, the sector's high demand generated inflationary processes affecting the cost of housing production, land, and construction materials. Following the second phase of the MCMV program, the federal government announced a target of 2 million dwellings for the 2011–2014 period, with an average of half a million units per year (USD 67.2 billion, including USD 28.6 billion in subsidies).

Colombia and the production of social housing schemes

To many observers, Colombia has been experiencing a housing bubble for the last 15 years, with housing prices rising 200% in nominal terms, and yet, the housing deficit has dropped from 56% to 25% (Arbeláez et al., 2011, p. 7). Even when only 46% of dwellers are homeowners, mortgage credits have grown by an annual average of 13%, accounting for 21% of GDP in the last decade, and along with Chile and Panama are the highest in the region.[7] While securitization has been associated with credit growth instead of an economy dominated by bond financing, this practice introduced a risk factor that may prove critical during housing crises. Nonetheless, Colombia has relied on banks using retail deposits as a primary source of financing and a general revision of funding schemes for mortgage lending has taken place, looking to reduce their reliance on short-term funding in favor of long-term savings that have the same rate of return as assets, reducing with it the exposure to interest-rate movements across mortgage institutions.

Drawing from housing finance history, Colombia created the *Instituto Nacional de Vivienda de Interés Social y Reforma Urbana* (INURBE) in the early 1990s to be in charge of housing policy, regulating and administering subsidies that targeted the low-income sector demand, as well as providing technical assistance to local entities and nonprofit housing associations. Later on, land policy regulations required municipalities to develop territorial plans with specific instruments in order to regulate land price increments that would not exclude low-income households from the housing market.

By the end of the 2000s, mortgage credit declined from a 6.2% share of GDP to 3.5%, funding less than a third of the total demand and leaving 70% of the informal market unserved, while accounting for 34.6% of the housing deficit, which was equivalent to nearly 4 million households of whom 12.4% lacked a house and 23.8% lived in inadequate dwellings (Fig. 4.13). The shortage could be traced back to the limited program of subsidies for low-income households and that partial credit guarantees as credit mechanisms that overcame the obstacles were tied to a lack of sufficient collateral to access a mortgage. Instead, loans increased 11.7% for households with higher levels of income,

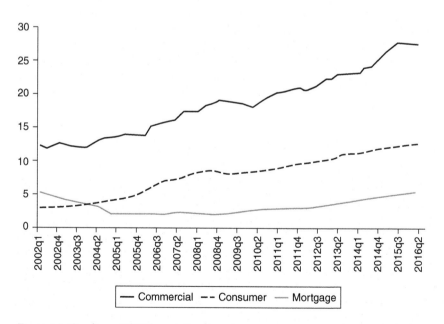

Fig. 4.13 Credit to GDP by market segment.
Source: Banco de la República, 2016.

subsidizing middle-income housing with a better standard of living, all of which increased segregation mechanisms and furthered inequality (Arbeláez et al., 2011). As we have suggested, housing policies did not solve *right to the city* demands, since 61.8% of the lowest-income quintile lacked access to sewerage systems and 40.4% lacked water networks, while in the second lowest quintile 43.7% lacked sewerage and 28.8% access to water networks.

In the early 1990s the Colombian government had a line of credit that funded housing, which was replaced by a multipurpose banking scheme that enhanced speculative mechanisms responding to capital market dynamics. At the end of that decade, a profound economic crisis took hold, affecting developers, banks, and mortgage debtors, triggering a governmental bailout, just as the subprime crisis of 2007–2009 developed. Colombia created a national entity that offered subsidized dwellings for low-income households with a long-term financial scheme using below-market interest, just as most countries in the region had (Fig. 4.14). At the time, partial mortgage guarantees were provided by the National Guarantee Fund (Fondo Nacional de Garantías) to finance intermediaries holding mortgage portfolios that had been previously evaluated according to risk-assessment procedures, and that could be made effective in the

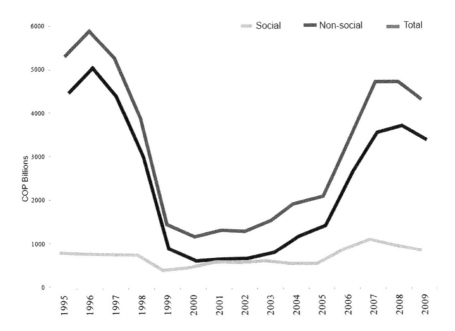

Fig. 4.14 Mortgage disbursements.
Source: Banco de la República, 2009.

event of an 18-month default or if the financed unit was given as part of a loan payment (the guarantee covered up to 70% of expected loss for individual loans with financial intermediaries).

With moderate success, these institutions reduced the housing deficit, admittedly supplemented by informal self-construction practices, and yet neoliberal reforms replaced these mechanisms with a system of direct subsidies on the demand side, but notably leaving behind the lower-income population, as well as informal workers who could not demonstrate their income sources. Since urban policy was oriented to supporting major development companies, the quality and size of dwellings declined, as did the location and urban infrastructure, deepening the mechanisms of segregation as well as adding further costs to peripheral housing developments.

Part of what drove these changes were the liberal reforms introduced during the 1990s, such as an indexing mechanism tied to inflation (known as UPAC) to guarantee investors a positive real return for their savings. This meant that, along with the privatization of banks, it was possible to offer positive returns and immediate liquidity, significantly increasing resources for construction and hedging capital market volatility. Large financial groups linked to mortgage banks captured these resources and partnered with developers that dominated the real estate market and imposed monopolistic prices on customers. With the introduction of multipurpose banking schemes,[8] the Banco Central started to shift its indexation from inflation to interbanking interest rates, making the sector more dependent and vulnerable to financial market fluctuations. Once again, the country introduced public policies that were intended to endorse private production of social housing rather than providing decent dwellings for the low-income population.

As with other countries in the region, Colombia adopted market-oriented neoliberal economic reforms to guide public policy. It eliminated the *Instituto de Crédito Territorial* (ICT) that once had been a mortgage originator and housing developer, only to hand it over to the private sector, which took control by creating new public entities such as the National Institute of Social Housing and Urban Reform/Instituto Nacional de Vivienda de Interés Social y Reforma Urbana (INURBE), which assumed the credit functions. The credit scheme consisted in the provision of a 10% down payment by the borrower, a 40% housing subsidy, and a 50% long-term conventional loan (Fig. 4.15). Along with those changes, privatization of public services, banks, and formerly State-owned entities produced great liquidity that later translated into high demand for properties, duplicating home equity loans, which produced a 40% growth (in real terms) in the construction sector's share of the GDP (DANE). However, after the liberation of financial markets in 2004, a sharp economic decline followed and housing prices plummeted, leaving housing development companies with large stocks of housing units sitting

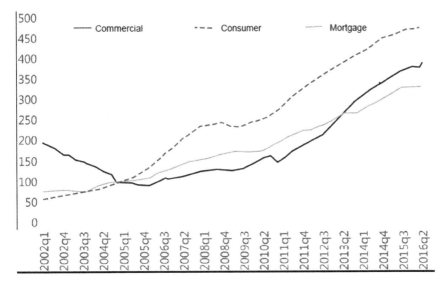

Fig. 4.15 Colombia. Credit by market segment (in real terms, Index, 2005Q1 = 100).

Source: Banco de la República, 2016.

on major land reserves, for which some paid back their debt with housing units or later filed for bankruptcy. As a result, land prices fell 48% and housing units declined in value in the process, significantly affecting the real estate market (Jaramillo & Cuervo, 2014, p. 47).

At this point, the central bank had to adjust its interest rates (20% real), which increased private developers' debt and ultimately impacted mortgage debtors' ability to pay. Moreover, the crisis led to massive layoffs, lower real wages, and increased unemployment. As in the subprime crisis in the US, the Colombian government bailed out banks while more than 70,000 households faced foreclosure. Moreover, the interest's indexation system shifted from inflation to commercial interest rates, but the High Court later ruled that this decision was illegal, forcing banks to give back mortgage payments based on the latter rather than the former.

That financial shock generated some structural changes in the system, linking the indexation to inflation and putting a cap on interest rates for social housing finance. The system of subsidies directed to the consumer proved to be complicated, since beneficiaries could not find housing that matched their budget. Nevertheless, in the last decade the Colombian economy has been growing and increasing its demand for housing, so the government chose the construction sector as an economic priority as well as an engine for development. The State has extended tax exemptions for major construction companies as well as for upper-income sectors to buy property, since the real estate market and the housing sector have countered unemployment effectively, even if those tax cuts also impacted government revenues.

Another example worth mentioning is the *100,000 free housing units program*, launched in 2013 by the housing ministry, which was 100% subsidized, although local authorities had to provide land banks for the projects as well as basic services (a considerable number since in the last decade the State built around 10,000 social housing units per year). The program targeted the lowest-income population, who had been displaced by paramilitary groups or violence in their hometowns, placing dwellers in peripheries, despite marginalizing neighborhoods due to their location.

Housing prices escalated during the 1990s, partially as a consequence of the structural adjustment policies that forced deregulation and stimulated leverage in the financial markets. However, by the end of the decade a major crisis arose due to increasing interest rates, which led to home foreclosures. As so often happens in such cycles, this was followed by a bailout of the financial institutions. More recently, over the last decade, real estate prices have once more been rising steadily in Colombian cities while real wages have largely remained stagnant. This situation is due to the use of financial mechanisms that have stimulated the demand, as well as the financialization, of secondary residential markets. This may create the conditions for a new housing bubble restraining the acquisition of living space in cities, and even exposing mortgages to default in the near future, partially due to the promotion of policies for property acquisition that distort market prices. In the meantime, as in other countries of the region, the production of housing developments in peripheral and underserved locations results in concentrated segregation, poverty, and violence, ending as a self-fulfilling prophecy.

Hyman Minsky (1982) claimed that financial crises were endemic to capitalism, and that they appeared when the excess optimism of economic prosperity generates a financial bubble that inevitably leads to a market crash. At some point, debt increases beyond the value of the asset through leverage, yielding profits out of limited capital due to the appreciation of prices created by others who seek to profit in this way. When debt-ridden capital reaches a threshold and financial entities lack sufficient funds, this accelerates the collapse of the whole structure of creditors, debtors, intermediaries, rating agencies, and even entire economies. Minsky's financial instability theory is framed as the result of excessive indebtedness following a period of apparent growth, a crisis in which some of the financial institutions, insurance companies, pension funds, hedge funds, and major investors benefit or may benefit by being bailed out in conspicuous circumstances. However, in the cases that have been examined, housing bubbles occurred partly when public policies endorsed financialization practices. According to Sommi (2005, p. 3), the nature of capital is to seek rent, flowing in larger volumes, intensity, and velocity to the most profitable sectors instead of to those that society needs most, for which financial crises are necessary for the reproduction of capital. In this sense, housing markets respond to an economic and financial rationale that affects the wealth and welfare of families, and therefore the unregulated function of those markets may increase inequality, segregation, and injustice in the production of urban space.

The housing sector serves as an important economic engine generating a number of spillovers, where housing prices respond to the high elasticity in construction costs, real interest rates, capital flows, demand, and income, determinants that tend to follow a cyclical pattern (Clavijo et al., 2004). In the case of Colombia, public subsidies and access to credit have not been high enough for families to access housing solutions, in the process maintaining a structural deficit that furthers inequality. In fact, Gómez-González et al. (2013) conclude that the main economic reason behind housing bubbles in Colombia has been that there was an increase in demand while supply plummeted, which later accelerated the reduction in mortgage interest rates. Previous to that, real estate prices had risen exponentially, creating expectations of even higher returns and leading to more demand that is independent of real wages stagnation, which is a common feature among the four cases.

In the last decade of the twentieth century, important changes in housing demand had been taking place in Colombia, where women accounted for 54% of the labor market, and a high percentage of middle- and high-income population are not homeowners (57%). This has been identified by the Colombian government as a financial niche for housing investment, launching housing acquisition programs such as "Mi casa ya" (My house now), which provides a 20% down payment subsidy to purchase a home plus a 4% interest rate subsidy on the mortgage loan each month, and yet the program has been questioned because it increases demand and brings distortions, affecting prices and creating volatility in the housing market (Ortalo-Magne & Rady, 1999). It is precisely the instability of housing prices that has been associated with economic bubbles, where crises grew deeper than financial recessions when resource mobility was compromised, since households spent a significant amount of their resources on their mortgage, preventing them from using their liquidity for other purposes. Also, homeownership – implied in a mortgage scheme – may prevent mobility to jobs, education, and other opportunities, which affects not only household welfare, but also the economy of cities.

New perspectives on housing policies in Latin America

A global trend to meet housing financial needs involves private pension funds as well as insurance companies that invest in government securities backed by assets such as mortgages. International banking institutions such as the World Bank and the International Development Bank have been endorsing a shift to such market-based financial systems in the region in order to converge toward international standards, stressing the claim "to deepen financial intermediation to meet the financial needs of under-served segments – in particular to increase housing finance," therefore transforming resources into long-term investments (Cheikhaouhou et al., 2007, p. xiv).

To this end, structured finance separates the pool of collateral assets from the credit risk of the originator, usually through special-purpose vehicles and using institutional pension funds that have limited capabil-ities to invest in what have been considered a risk-free asset, holding as it does the highest domestic credit rating (AAA). In order to widen the supply of credit-worthy bonds on domestic financial markets, structured finance originators package pools of underlying assets that generate cash flow, which are turned into marketable securities and sold to investors, in the process enhancing domestic markets as well as creating a mecha-nism to spread risk among originators, institutional investors, and credit enhancers (Fig. 4.16).

For mortgage securitization purposes, it is important to standardize lending documents and underwriting procedures, as well as professional criteria for property appraisal, which protect investors from the bankruptcy of originators or servicers. In this sense, governments, professionals, and regulators should agree on procedures to guarantee the quality of the assets backing the securities traded in secondary markets through strong legal and regulatory frameworks and sound financial markets. Besides economic upheavals, in the last two decades Mexico and Colombia have achieved sustained growth in the region while maintaining relative macroeconomic

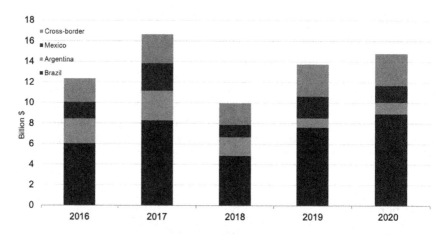

Fig. 4.16 Market-wide Latin American structured finance issuance.
Source: Standard & Poor's Financial Services LLC, 2020.

stability, which has enabled them to develop mortgage securities markets, while Chile met housing finance needs with mortgage bond issuances. But the systematization of mortgage portfolios in Latin America still needs to be fully developed, as even when its share of the national GDP reached 11% in Chile and Mexico, in Europe and North America it accounts for an average of 40% of the GDP.

Mortgage securities in the region have emerged during periods of significant economic growth or as a countercyclical measure after financial crises. This was the case in Brazil and Colombia, where legislation was passed in 1998 to allow mortgage securitization, including the creation of *Titularizadora Colombiana*, a private entity capable of buying portfolios from banks, packaging them into mortgage-backed securities, and using them as collateral, in the process providing liquidity to the market as well as diversifying risk for creditors.[9] This company has securitized about 30% of all outstanding mortgages since its creation, with approximately half of the bonds sold to insurance companies, financial entities, and, to a lesser degree, pension funds (Titularizadora Colombiana, 2005).

In the case of Mexico, insurance companies, pension funds, and even Chilean insurance companies purchased mortgage-backed securities, while the *Sociedad Hipotecaria Federal* targeted non-bank financial entities (Financial Societies of Limited Purpose, SOFOLES) that provided mortgages to middle-income households. These entities were not permitted to take deposits, which proved to be a major liability during the early 2010s crisis. Therefore, Chile, Colombia, and Mexico allowed pension funds and insurance companies to invest in mortgage-related securities as long-term assets, and in the latter case, private pension funds invested in MBSs because interest was granted a tax exemption, which Colombia's institutional pension fund did not benefit from.

Housing is a constitutional right in most Latin American countries, and there are political and social movements for housing rights that advocate for tenant protection and against eviction procedures, yet capital markets as well as institutional banks have required Latin American States to enforce property rights, putting in place foreclosure procedures in their legal systems, leading Colombia and Mexico to pass laws to reduce foreclosure times from five to two years on average. Nonetheless, a complicated balance arises between the protection of consumer and creditor rights.

One of the keys for the securitization of mortgages is that a special-purpose vehicle separates the pool of collateral from the issuer or servicer (in order to obtain off-balance-sheet accounting), and works as a subsidiary created by a parent company to isolate financial risk in the case of a major financial crisis. When governments endorse financial and housing markets they ought to put into place adequate laws, taxes, and regulations that enhance mortgage securities, tailoring loan products for consumers, mitigating liquidity, cash flows, and credit risks (Fig. 4.17). Latin American countries learned that this practice will always have

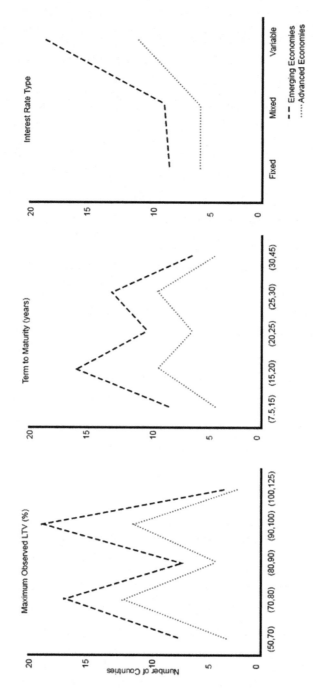

Fig. 4.17 Cross-country housing finance characteristics. (LTV or loan-to-value ratio is an assessment of lending risk before approving a loan).

Source: Cerutti et al., 2015.

a downside, since whenever governments stimulate mortgage capital markets it always comes at a cost, where risks are transferred to the State at the time when they provide financial guarantees to be rated as sound investments, sometimes needing credit enhancements that warrant timely payments by second-tier banks.

Notes

1 These schemes have been operating in Mexico since the early 2000s and, due to particular circumstances, their operation had resulted in 6.1 million vacant social housing stock across the country in 2020.
2 RMBSs were modeled as BORHIS (SHF), CEDEVIS (INFONAVIT), TFOVIE (FOVISSSTE), CEBURS (Certificados Bursátiles).
3 World Bank national accounts data, and OECD National Accounts data files. *DP per capita (current US$) – Brazil*. https://data.worldbank.org/indicator/ NY.GDP.PCAP.CD?end=2019&locations=BR&start=2000.
4 https://www.ibge.gov.br/estatisticas/sociais/habitacao.html.
5 Group 1: 3 MMW (monthly minimum wages), equivalent to USD 754; Group 2: 3–6 MMW (USD 754–1,508); Group 3: 6–10 MMW (USD 1,508–2,513).
6 This was the way the Brazilian system operated before the changes to their legislation, which then allowed the rapid financialization of the real estate market.
7 Mexico 10%; Brazil 9%; see www.hofinet.org and Central Banks.
8 This type of banking provides financial capabilities and investment strategies, and it is authorized to carry out all those financial operations including investment banks, monetary market funds, etc.
9 Titularizadora Colombiana is owned by mortgage lenders and the International Finance Corporation, a private company that is a lending arm of the World Bank, with no support from the State.

5   Financing megaprojects in Latin America

Megaprojects: finance, markets, and the State

Megaprojects are large urban projects, which are particularly challenging due to the complexity of their management. There is a straightforward definition for the term as it refers to a wide range of interventions, from corporate business districts, tourist resorts, and large residential developments in the peripheries of cities, to the regeneration of historic centers, or the recycling of deserted public land within the city into industrial zones, airports, train stations, etc. These large projects have a substantial impact on the territory, transforming its economic, social, and political dynamics. In what follows we examine four interventions in the Latin American region, each with a different framing and objectives, scale, and organization.

The idea of a megaproject emerged in Europe during the 1970s as a strategy to solve the contradictions between urban planning and large architectural projects, providing it with a framework to integrate elements of spatial planning along with multidimensional proposals that considered the socioeconomic conditions to develop as part of the general layout of the city. From this perspective, large urban projects were considered growth engines that could lead to strong centralities, while integrating the private sector into the economic equation through the use of incentives such as zoning regulations, tax exemptions, etc., or building large public infrastructure projects to enhance land values. According to Borja (2001), large urban operations should serve to address several urban problems at once, including the economic, social, and spatial realms through a series of actions following a single rationale and corresponding to an overarching plan. Unfortunately, the majority of large urban developments in Latin America have lacked participation mechanisms to guarantee the inclusion of mixed-income populations and, instead, interventions have mostly followed the staging of major redevelopment projects with plenty of media exposure but little social consensus, where culture has been framed as an industry capable of attracting financial investments. As a result, these plans brought about growing opposition and social responses

DOI: 10.1201/9781003119340-6

in recent decades from people who sought to resist their potential for environmental destruction as well as gentrification and segregation.

As participants in megaprojects, financial-market stakeholders have claimed that they require the renovation of regulatory instruments as well as a different planning framework that allows the integration of large urban projects into the general layout of the city, while supporting the introduction of exceptional measures to endorse these urban interventions. The rationale behind these projects is that the public and the private sectors are sharing the risks, and yet with these partnerships the projects may be able to bypass urban and zoning regulations, in the process curtailing mechanisms of participation, and imposing new governance procedures (Swyngedouw et al., 2002).

Megaprojects take place at different temporal and spatial scales, apart from dealing with complexity and uncertainty among public and private actors. They have been instrumental in attracting local and international capital as well as serving as financial conduits for these funds, even as they entail long-term investments and present a degree of uncertainty during economic and/or real estate crises. Moreover, large urban developments are rarely part of comprehensive urban plans, as they do not include rigorous analysis of the project's impact on land markets and often replicate social divisions and territorial inequalities.

But these interventions have not yet developed mechanisms to capture the capital gains that could be used to compensate for the externalities of the projects or to prevent further inequalities. Instead, planning instruments, including tax exemptions, density bonuses, and regulations, are used to attract private investments through structured finance to achieve specific goals, in the process diverting public money and capital gains for private ends. To guarantee the public interest, large urban projects need to develop a complex framework that meets the ensemble of actors and interests for whom the State serves as facilitator or manager and who will not necessarily favor the private partners of the operation. In this sense, the public sector draws investments from the private sector but assumes the costs of subsidies, tax breaks, and exemptions. Overall, these projects should be framed in a way in which all actors will benefit from capital gains and increases in the value of real estate. Since capital is not guaranteed by the public sector, these partnerships have to deal with risk through different frameworks of operation such as trusts, corporations, or public-private partnerships.

The role of the public sector in these projects is central, leading the first investments in infrastructure as a public endorsement for the project, and even serving as a guarantor for the initial financial operations. The contradictions and paradoxes emerge when urban planning is privatized, public responsibilities are disregarded, and interventions are concentrated in high-return areas (Rodriguez, 1995). Since the projects have a considerable impact on the urban configuration, the private sector investments in the area compete with the city, one might say, creating special governing entities in charge of planning and building the sites, often managing, taxing, and maintaining the ensemble.

The projects may attract major public and private investments, and yet they can result in isolated fragments of the city, featuring tailor-made regulations and barely any public participation. Sometimes the projects include social and middle-income housing areas, but as developments continue to raise land values, revisions and amendments end up eliminating such provisions. In most cases, land value appreciation creates further differences and inequalities, since capital gains benefit investors and create gentrification mechanisms that expel local residents as a *natural* device for the concentration of capital (Valenzuela Aguilera, 2013). To get a better sense of how these programs work and their consequences, let us turn to a series of examples.

Puerto Madero, Buenos Aires

Major urban operations developed at the end of the twentieth century in cities such as London, Boston, Baltimore, and Barcelona, and in Latin America projects included Faria Lima (São Paolo), Santa Fe (CDMX), Porto Maravilha (Rio de Janeiro), Malecón 2000 (Guayaquil), Puerto Madero (Buenos Aires), and Punta Pacífica (Panama). These were significant interventions that transformed land values, modified land uses, and introduced new management mechanisms and governance structures. In general, firms undertook them where the realtors and international developers were one and the same.

After several attempts, the Corporación Antiguo Puerto Madero (CAPMSA) was created in 1989 as an autonomous management entity to coordinate public and private actors and advocate for the public interest first, and later handing over the decision power to private investors. The idea of revitalizing the docklands and regenerating them was a powerful commitment, as was the intention of restructuring the area through the construction of public spaces, residential compounds, and better infrastructure, and yet these benefits were intended to increase the area's land value, cost opportunities, and global allure. These objectives changed the project into a high-profile real estate development, one that stood in contrast with the rest of the city and displayed the social and economic differences that mark most Latin American cities.

As with other projects designed to gain visibility in the global arena, Puerto Madero attracted local and international investments, creating exclusive urban enclaves that belonged to a global network of high-profile real estate markets (Harvey, 1989a). This, and similar strategies elsewhere, have favored particular projects for certain social and economic interests, particularly sports facilities, shopping malls, business districts, or research centers, while leaving behind social housing developments, public schools, public health facilities, or public transport systems. Even when urban interventions raise the tax base for public expenditure, profits for the investors generally exceed contributions, and in the present case, through their "semiautonomous status" the promoters were able to attract both a talented workforce and real estate investments, despite the initial intervention resulting in large amounts of vacant residential, office, and commercial real estate.

Even when redevelopment projects in the 1980s and 1990s were part of an economic revitalization strategy, they transitioned to the financialization of real estate after a few years. Large urban projects like Puerto Madero follow a trajectory: they increase land values from the announcement of the project; followed by the construction of the first – probably iconic – buildings that will provide initial securities for investors; and later selling land already served with public infrastructure; followed by commercializing land and properties that have already produced the yields from land appreciation and capital gains mechanisms, which are part of a speculative strategy to capture land value increments. Also, partnership with the local authorities allows developers to capture capital gains from the provision of public infrastructure, as well as changes in land use and density that will impact the real estate market, and that are not always accounted for.

Puerto Madero dates back to the second founding of Buenos Aires in 1580 by Juan de Garay, who sought to open a direct connection to Alto Perú (now Bolivia), and established the Puerto de Santa María de los Buenos Aires. The city later became the capital of Río de la Plata Viceroyalty. The commission to build the port was granted in 1887 and it was completed in 1898 by engineer Eduardo Madero – in association with a British company, also building the Costanera Avenue and the municipal water park in 1916. However, Puerto Nuevo was built a few years later, rendering Puerto Madero obsolete for most of the twentieth century. The governing entity for Puerto Madero was ruled by Corporate Law, meaning that it was not compelled to share information on its asset management and that it was only audited by private accounting firms, having various competences and few public responsibilities. These entrepreneurial urban policies moved the project away from the public interest, including generating jobs and increasing the general welfare. Nevertheless, at the end of the operative stage, profits accounted for USD 150 million, which were divided between the public and private partners (Garay, 2002, p. 8). The latter used it for housing (50%), education (23%), and environment (2%) (Boletín Oficial de la Ciudad de Buenos Aires, O:M.44.945, 1991). At the end of the day, the promise of capturing capital gains as an instrument to share the profits with the general public remained an ongoing subject of debate given that the public sector has to adopt a speculative rationale when acting as a developer, creating value, while controlling or interfering with the market's dynamics (Fig. 5.1).

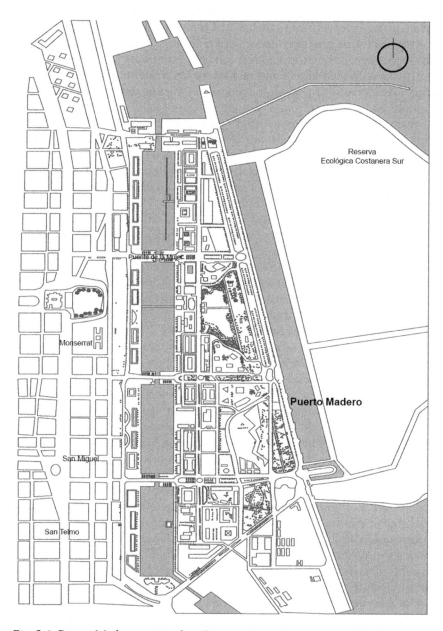

Fig. 5.1 Puerto Madero masterplan, Buenos Aires.
Source: Etulain, 2008/CAPMSA, 1992.

A century later, Argentina experienced major processes of privatization, deregulation, and economic liberalization during the 1990s, which attracted global financial capital, in the process exacerbating inequality and social polarization. At the time, the State invested heavily in infrastructure, undertaking the construction of highways to link metropolitan areas, while private investors built shopping centers, residential complexes (known as *Countries*), high-end hotels, and business districts (Ciccolella, 1999, p. 15). Among those projects, Puerto Madero was envisioned as the poster child of a large urban operation to regenerate an area of strategic and historical value for the city.

Located in the oldest sector of the waterfront district, the Puerto Madero project covered 320 acres of land, and nearly 100 acres of water, including four dikes and 16 docks. As part of an economic strategy, former President Carlos Menem (1989–1999) endorsed the project, and two former State companies merged creating Corporación de Antiguo Puerto Madero (CAPMSA), to be managed by national and municipal partners, and later joined by local and international developers.[1] First, the corporation hired a Catalonian consulting firm led by Joan Busquets, Jordi Borja, and Joan Alemany to design a strategic plan to transform the area into a new waterfront, integrating the Costanera as part of the city's urban green system (Borja & Castells, 1997b, p. 353). The proposal was intended to perform a kind of urban acupuncture where interventions would be capable of producing different and desirable outcomes, activating a chain effect. However, large urban operations of that scale had to be framed as a public initiative, even under a public-private structure, and seeking a middle ground between the corporation's interests and the government which did not necessarily meet citizens' expectations and instead yielded toward the economic benefits the project would provide.

The Catalonian team set up *Tecnologías Urbanas de Barcelona, Sociedad Anónima* (TUBSA) a public-private enterprise created to export urban technologies worldwide. The corporation offered all kinds of services, including projects, urban and architectural design, training and technology, implementation, and a professional practice that had already secured contracts in major cities in Spain, Latin America, and Northern Africa. They were the type of firm described above, whose projects targeted high-income developers to finance the operations, attracting private investments, and yet alienated the local population since the projects deepened the existing spatial fragmentation and social segregation.

The project intended to revitalize the area and develop an urban hub to enhance its centrality and reverse the process of abandonment and decay of the previous decades. The federal government created CAPMSA and transferred the land to the corporation, which proceeded to define a self-financed model, market the land, and supervise the masterplan. Apparently the project went well and after a couple of decades, CAPMSA was awarded several prizes and was featured as a best practice by the

UN-Habitat agency as well as other entities. Nevertheless, the operation has been criticized because of its exclusionary and divisive land-use practices, since they favored particular cultural, leisure, and sports activities directed primarily at the upper-class residents. It is noteworthy that the national government undertook this large urban operation in the middle of a major economic crisis with the intention of creating wealth as well as jobs, and to this end the Puerto Madero project attracted over 5,000 new residents and created more than 45,000 jobs (Garay et al., 2013, p. 10).

The project took place over several stages, starting in 1991 with a national competition that selected a team to design the preliminary urban project for Puerto Madero. The plan had a 20-year horizon and combined leisure with cultural activities, services, restaurants, shopping, workshops, and storage facilities. The project also contemplated green spaces such as a metropolitan central park, as well as transforming the Costanera Sur into an ecological reserve.[2] A decade later, the project supported the turbulence of a major economic crisis and, after the 2003 elections, Argentina restructured its foreign debt and managed to overcome the turmoil – which brought a new impulse to the real estate market, enhancing the value of centrality for new investments. By 2011, CAPMSA had sold USD 25.7 million in real estate and invested USD 205 million in infrastructure and management, even incorporating 100 acres of water bodies and 70 acres of parks. The following year Puerto Madero extended to the south as part of the creation of an arts district.

Puerto Madero brought in real estate investments of close to USD 2.5 billion (USD 6 billion actual value), which impacted tax revenues significantly (it is estimated an annual property tax worth USD 12.4 million). The project boosted the labor market, creating 6,260 construction jobs over 20 years, 27,000 administrative jobs, and 45,000 jobs in the service industry. Nonetheless, social outcomes were not as visible, since the tendency toward enclosure prevailed in the upper classes, creating vertical *gated* communities, isolating themselves from the rest of the population. Even though social housing was part of the original proposal, land values eventually expelled the low- and middle-income population from the project grounds. Being a public entity, the corporation chose how to distribute their dividends to shareholders, for instance, being able to invest its profits from land sales in services and infrastructure that would later raise the value of properties. However, at the end of the day, profits from capital gains benefited their major investors instead of the public sector, since they were not contemplated in the initial project through value-capture instruments to finance social housing development and other government programs (Etulain, 2008).

In the case of Puerto Madero, the public-private partnership benefited the local tax distribution, but brought major yields to investors, where the government served as a mediator to provide the best financial conditions while hedging risk. As a control provision, buyers could only claim their land titles once their projects had reached 40% of their completion (CAPMSA, 1999, p. 73). During the construction process, CAPMSA had to deal with local and international developers in order to attain its urban and social targets. In retrospect, the project created new green spaces for the population but did not include social housing developments or foster social equilibrium since the plan was never intended to be comprehensive, despite its innovative financial and management instruments. Moreover, public land was sold to private investors and upper-income buyers, while public revenues were used to improve services and amenities that benefited residents and increased land values even further. The local investors included politicians and entrepreneurs linked to money-laundering scandals that frequently had their headquarters in the area and profited from financial instruments to invest their illicit money (Lawrence, 2017).

In this respect, Harvey (1989a) suggests that in these circumstances a scheme of urban entrepreneurship takes place, characterized by the use of local power to attract private investment, in which the public sector assumes the risks and the private sector receives most of the benefits where urban projects follow a speculative logic that furthers territorial development. This new framework was designed to counteract the economic slowdown of the 1980s by reappraising land value in a real estate market intended for international investors who were able to position development projects in the global arena. However, this business rationale clashed with social actors' interests that did not partake of the benefits of the project, and instead developed into social struggles, generating coalitions and class alliances, contesting that the local government favored the interests of capital.

It is typical for these projects to be considered an exception to standard zoning requirements and usually require changes in the existing planning ordinances and zoning provisions, which are tailored to the developers' needs on a case-by-case basis. Ultimately, what is at stake is the right to the city, which stands at the crossroads of political and social powers' interests for the appropriation of space and which, in essence, implies the resolution of conflicts (Rolnik, 2015). In this battle, the most vulnerable population is displaced to less desirable locations in the city through gentrification as a mechanism resembling social reengineering.

It is important to underline the role of the State in major urban operations, adapting codes and legislation, while in the case of Puerto Madero, after the corporation rehabilitated and sold the docks, the local government used the revenues to build infrastructure, public spaces, roads, and street lighting, while favoring certain land uses that helped expand and increase the central city's facilities. The experience with large urban projects is one of mixed results, where speculation, the transgression in the use of urban instruments, the lack of transparency, and furthering social segregation and spatial fragmentation is not rare. Also, these projects need to enhance spaces for participatory planning, social housing, and public facilities, which have not improved over the years, disregarding the possibility of using revenues to benefit other social groups through redistributive mechanisms, and instead remain as isolated and disconnected enclaves.

Santa Fe, Mexico City

In Mexico Santa Fe was one of several megaprojects launched in the 1990s intended to position the country in the global arena as having a strong real estate development industry. At the time, other large urban projects included the regeneration of Mexico City's historic district; the Alameda project, meant to convert the Alameda park into a local Central Park; the conversion of a top-end residential neighborhood into a shopping district for design and fashion (Polanco); and the Xochimilco park regeneration project that was intended to accommodate entertainment industries. Always looking to attract private investors, Santa Fe became the location of more than 2,000 corporations; it was envisioned as a self-contained city that included malls, residences, condominiums, restaurants, private schools and universities, hospitals, and sports centers. At some point, the Residents Association was created and convened with the local government to allocate part of the property tax revenues for the maintenance of the infrastructure, the provision of urban services, as well as private security.

Santa Fe is located in the western part of Mexico City that was heralded as part of an environmental strategy to create a new centrality for the city. The project included the restoration of woodlands for groundwater recharge and also served as a global enclave for multinational finance and real estate. Originally a dumping ground surrounded by sand mines and irregular settlements, the area was adjacent to high-profile neighborhoods such as Bosques de las Lomas, La Herradura, and Tecamachalco. Considered a large urban project, Santa Fe assembled public and private actors as political and economic elites mixed with key actors in the construction business and real estate markets.

During the 1980s Santa Fe represented a transition area between the expansive urbanization heading west and a conservation district in the upper part of the city, covering about 2,303 acres of sand mines and forestlands. This enclave was conceived from the beginning as a globalization device (Valenzuela Aguilera, 2013) to connect the local economy to capital markets, real estate ventures, and business services through a public-private entity. This managerial innovation allowed the project to overcome the apparent conflict derived from being situated between two different administrative units (Cuajimalpa and Álvaro Obregón mayoralties), without having to create a different independent political entity. The project reproduced the prevailing socioeconomic division in the city, where wealthy areas contrasted with adjacent communities in terms of income, unemployment, or educational level. This social stratification has been acute, where the upper end dwells in residential complexes with private security, common spaces, sports facilities, and commercial and office services, while the bottom end lives in informal settlements, working in services, maintenance, and the informal sector.

Santa Fe's marketing strategy envisioned the project as part of an environmental proposal to transform a former garbage dump into an "ecological" real estate development (Gaceta Oficial, 2000). Framed as part of a "green revolution" to recycle garbage and increase the natural reserves in the city, the masterplan proposed a dense high-rise skyline, and although 34% of the project was destined for green spaces and ecological reserves, these areas were inaccessible ravines that cut across the territory, a park built on top of the garbage dump (which never opened due to contaminated soil), a private golf course, and a public park which later was transformed into a shopping mall.

Although large-scale urban projects in general follow a particular political agenda, Santa Fe transitioned through several different governing structures and political affiliations, as will be discussed below. But even when aiming at different objectives, the State power structures were seldom monolithic, which allowed the continuation of these projects (after all, they provided valuable revenues to local governments). Of course, political groups often frame such projects with economic, environmental, or global discourses, pleading a status of exception for the area to allow special regulations, tax exemptions, special provisions, zoning regulations, and self-governance structures in the name of the greater good, and Santa Fe was no exception (see Fig. 5.2).

The first administrative provision was to name Santa Fe as a Special Controlled Development Zone (ZEDEC), a designation that allowed the application of specific regulations for that sector of the city that was not part of the city's metropolitan planning code. Under these procedures, 31% of the surface was designated as an ecological conservation area, 25% for roads and highways, 15% for housing, and 27% for residential, commercial, and office use. Later, the ZEDEC was transformed into the Urban Development Program of Santa Fe that, along with another entity called Polygons of Action, allowed the transformation of land uses as well as further subdivision (as in the case of La Mexicana estate, where 8,250 social housing units were initially granted). Also, among other initiatives, the program contemplated the *disincorporation* of three irregular settlements to be assigned to the Álvaro Obregon mayoralty (being now responsible to provide them with infrastructure and services), as well as maintaining considerable areas of land for conservation uses, but that ultimately benefited real estate interests.

To guarantee the latter, the ZEDEC Santa Fe Neighborhood Association was created in 1994, assembling representatives of national

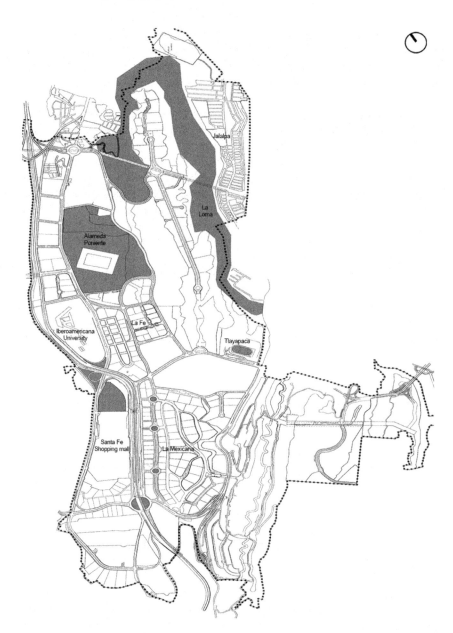

Fig. 5.2 Santa Fe masterplan, Ciudad de México.
Source: Gaceta Oficial del Distrito Federal, 2012.

and multinational corporations, as well as educational, technological, and financial companies, but no community groups. The association played an important role in the management of resources and private security, along with Servicios Metropolitanos S. A. de C.V. (SERVIMET), a public company that supervised the operation of the urban project as a whole. Later, as of 2014, a public-private trust was created for the same purpose, with an administrative board that included four members of the association and three representatives of the city government (both delegates of the local mayoralties were invited), but again excluding any sort of social organization.

This new administration carried out urban upgrading projects in the informal settlements in the north of Santa Fe, which were financed by capital gains capture mechanisms, that is, they originated whenever public interventions in the area raised land values that were later retrieved through a special tax. In recent years, Santa Fe has been the scene of various social conflicts, like the protests over the construction of the Supervía Poniente Highway that runs across informal neighborhoods, causing environmental damage (Pérez Negrete, 2017). Also, the construction of Garden Santa Fe shopping center on the single area that had been designated as a public park caused much upheaval, as did the cancellation of a social housing development in La Mexicana estate, and residential towers could be built there instead. Several of these actions took place during a series of neoliberal administrations that favored business and speculative interests in that area.

The case of La Mexicana estate illustrates the conflict that comes from the clash of interests when an elite is mobilized to prevent the construction of a social housing project in the area and instead supported the construction of condominiums and a public park. To this end, using an instrument called *activation by cooperation*, a real estate investment trust (FIBRA DANHOS) and a developer (Grupo COPRI) agreed to build the park in exchange for 30% of the property to build and sell 1,650 luxury apartments. This kind of project has served as an instrument to legitimize the neoliberal urban model, increase social inequalities, socio-spatial fragmentation, and the concentration of capital, thus creating alternative realities for the rest of the population.

To carry out such megaprojects, the establishment of appropriate executive agencies is required to meet financial objectives, meet market trends and variations, and comply with the social requirements to which each project is committed. However, the latter generally does not come to fruition, and in the best of cases it is likely to be only partially fulfilled: since the operational structure of the project did not include participatory mechanisms, a disconnection between the operational group and the inhabitants will always exist.

According to one of Santa Fe's advocates, the project was part of a more ambitious plan to restructure the Metropolitan Area of the Valley of Mexico (González & Martínez Almazán, 2018), with the intention of creating a new centrality in the west of the city to reduce the need for intra-urban displacements while achieving economic self-sufficiency in the area. This regional and polycentric model aimed to consolidate urban sectors dedicated to services where jobs would be concentrated near the so-called strategic territorial reserves. To this end, the donation of land to locate a private university had among its main advantages the positioning of a high-profile segment within the real estate market, partnering with local government to build access roads, as well as promoting the closure of the open-air dump because it was located in the vicinity of educational facilities. To allow real estate development, some of the so-called *irregular settlements* were relocated and the land was paid for with the same sand that was extracted from the mines. Later, a hydraulic plan was designed (drinking water networks, collection of sewage, gray-, and rainwater) that enabled the use of property taxes and water service rights as the main source of the project's expenditures.

Also, a system of highways across the project was built, but the connections with the existing public transport network were overlooked, preventing the construction of a bus rapid transit (BRT) line or the extension of the Observatorio metro line. These omissions were not justified for technical reasons but were the result of opposition from the business class and residents. They wanted to maintain the exclusivity of the area at any price, arguing for the importance of the ecological rehabilitation of the degraded mining area, in the process avoiding the location of poor informal settlements in the area.

The urban project was commissioned by the celebrated architects Ricardo Legorreta, Teodoro González de León, and Abraham Zabludowsky, who centered the main features of the masterplan design on vehicular circulation rather than pedestrian facilities or the design of public spaces for social interaction. The project profited from the relocation of the Iberoamericana University – which had been recently affected by an earthquake – a private education facility that would enhance the real estate market. In addition, a series of tunnels were built to connect the project with the Bosques de las Lomas residential cluster as well as with its affluent surroundings, which had the highest real estate value in the city. The mayor of the city, Manuel Camacho Solís, profited from the fact that Mexico was about to sign the North American Free Trade Agreement (NAFTA) with the United States and Canada, as well as the General Agreement on Tariffs and Trade (GATT), both of which would give the country a proper platform to interact with global markets through the third circuit of the space economy.

It is noteworthy that in the previous two decades, the completion of different projects had not been interrupted despite the political tendencies of the city administrations, including new infrastructure projects such as the Bridge of the Poets, the Supervía Poniente, the Observatorio-Toluca train, and La Mexicana Park. Real estate projects have a major appeal to different political affiliations since land appreciation and the associated speculation mechanisms generate resources and investments that convey political and social dividends. In addition, projects of this scale are justified on the grounds of generating value that would later return to the community, even though this has rarely been achieved. Likewise, large projects are usually implemented through special management entities that allow them to disassociate themselves from the obligations of public administration such as decentralized organizations, development corporations, or public companies, introducing a rationale focused on gaining profits rather than meeting social requirements. The apparent independence has allowed governing entities to make decisions swiftly and even negotiate with private partners without the intervention of social actors, as well as having a privileged position in obtaining permits and authorizations promptly. So, although these entities, as instruments of public interest, should respond to social purposes, their very structure is oriented toward profitability rather than the common good, for which the existence of reliable transparency mechanisms is mandatory.

In the case of Santa Fe, SERVIMET was created as a public company with a board of directors chaired by the mayor of the city and including top-level officials from his administration. The company was designed to be self-sustaining and to ensure that the project was carried out in accordance with the plans, policies, and programs of the city government, thus guaranteeing proper coordination between the different entities. For the operation of the project, the company hired specialists on a temporary basis to intervene in the different parts of the process. The project was financed with public-private investments, and only recently financial instruments, loans backed by mortgages and derivatives, are being considered as an option. However, an interesting mechanism introduced later was the Improvement Contribution Tax, which captured capital gains derived from the construction of public works that were funded with fiscal resources (such as infrastructure and facilities), as an alternative source of revenues.

Among the advantages of Santa Fe was the proximity to the supply lines of drinkable water for the city, as well as high-resistance ground that allowed high-rise construction. However, one of the arguments for carrying out the project was to reduce the displacement of people from their place of residence to their workplace, even if some years later only 11% of the 40,000 inhabitants work in the area (González & Martínez Almazán, 2018). Likewise, the project had obvious shortcomings, since mobility is compromised during rush hours, in addition to the fact that affordable and middle-income housing became an unattainable objective in the face of land prices, which compete with international real estate markets. This also diverges from the limited supply of groceries for daily consumption, mid-level restaurants, public health systems, or the existence of an efficient, economical, and safe public transport system. It is paradoxical that the masterplan features mobility at the center of the urban project, but it did not comply with the users' needs.

Large-scale urban projects' claims of financial autonomy are based on public-private partnerships, and to that end Santa Fe used structured finance debt instruments such as stock certificates, asset-backed securities, REITs, pledge of future credits and security agreements, financial guarantee insurance, etc. The improvement contributions tax allowed the public sector to capture the increase in the value of land produced due to the construction of public works, seeking with it a progressive redistribution of income as well as an important source of fiscal revenues.

Santa Fe stands as a parallel realm, encapsulated and distant from a contrasting reality that does not find proper linkages with the rest of the city, contributing to neither the territorial nor socioeconomic integration of the metropolitan area. Also, citizen participation does not seem to have a place in this kind of framework, since there is a missing link between its governing powers and the population, which is left behind in the decision making on the present and future of the area. It is paradigmatic that the Residents Association brought together representatives of corporations and other institutions but not the social organizations that extend beyond the area of high-rise buildings, particularly the middle- and lower-income neighborhoods. Santa Fe reveals that social participation without adequate and inclusive mechanisms cannot translate into any kind of governance or consensus building.

Punta Pacífica, Panama City

Panama City was originally founded in 1519 by the Spanish conquistadores and later rebuilt in 1673, growing inward for the next centuries. During the Spanish and Portuguese colonial period ports were strategic locations for commerce, trade, innovation, industry, and defense. In the seventeenth century, mining expansion in the country intensified the trade of slaves, locating its activities near Praça XV and later transferring to Valongo Wharf when the Portuguese royal family escaped the Napoleonic wars and settled into the Imperial Palace near that site (Cardoso et al., 1987; Lima et al., 2016).

Along with the international slave trade, the port developed trading counters and built warehouses, becoming the main economic engine of the region. In the following centuries, three working-class neighborhoods developed around the port activities, diversifying the kind of infrastructure that was needed, such as storage, warehouses, shipyards, and industrial and commercial sites. However, at the end of the slave trade period, as port-related activities declined, the area remained as a low-income township known for gambling and prostitution, stigmatized by the municipal authorities for its unsanitary, deviant, and marginalized conditions (Cardoso et al., 1987; Benchimol, 1992). As has been a constant in history, in the early twentieth century hygiene reforms after earlier pandemics resulted in urban interventions to widen avenues, eviction of low-income residents, and the imposition of new construction regulations in order to appreciate the real estate value of the area while vindicating it as the urban renaissance of that part of the city.

In the last three decades, Panama has had one of the highest rates of GDP growth in Latin America. Its economic stability has been associated with the presence of the US for over a century; its currency has been indexed to the US dollar since the construction of the Panama Canal back in the early twentieth century, positioning the country as an international trade and financial center. The indefinite concession to build, use, administer, and protect the canal served the emerging nation as a guarantee of US protection against Colombia, gained with its independence in 1903. Through the Carter-Torrijos treaty in 1997, Panama fully recovered its sovereignty over the canal at the turn of the century.

The canal provides steady employment and business opportunities that are associated with trade and services, which profits both locals and migrants. As in other Latin American countries, Panama established a constitutional democracy that was later dominated by a market-oriented oligarchy, which was replaced by a military junta in the late 1960s, until the US invaded the country two decades later. Allegedly, the intention was to overthrow the military regime of General Manuel Noriega, who was accused of drug dealing, money laundering, espionage, etc., even when he had been paid as an informant by the CIA for a decade (Lawrence, 2017).

Since then, the country has become a financial hub in the region, attracting major investments and increasing its GDP, in the process consistently positioning it as an emerging market. The country has been restructuring its economy to fit the market-oriented global trade, which has led to exponential growth of its real estate market. Despite its economic success, Panama has not solved the problem of the unequal distribution of wealth, holding one of the highest inequality rates in the region (OECD, 2019),[3] with 7% unemployment and 45% of jobs in the informal sector, while other areas such as education, health, governance, and environment have been left behind. However, according to a national survey on quality of life, 36% of the population are situated below the poverty line, while 16.6% are under the extreme poverty line (MEF/BM, 2008).

The National Institute of Statistics and Censuses (INEC, 2015) registers that the construction and real estate sectors accounted for 17.4% of GDP, while transportation, storage, and communications generated 25.1%, financial intermediation covered 7.8%, and business, real estate, and rental activities generated 9.7% of GDP. Therefore, the economy is dominated by the service sector, accounting for 80% of productive activities, where the real estate market plays a significant part, including residential tourism and resort markets, which account for 80% of the new developments. Failing to establish urban regulations to diversify its nodal centers, local governments favored densification strategies that later turned into high-rise real estate investments. Among this sector, luxury high-rise condominiums and beach villas in gated communities have spread all over the city, most of which are built as second homes or as speculative investments (Ganster, 2001).

Residential tourism refers to an emerging subsector where retirees buy or rent properties to spend extended periods of time, along with other real estate investors from the US, Canada, Europe, or neighboring countries such as Colombia and Venezuela. From these last countries, affluent businessmen invest in safer real estate properties, not necessarily to move into or rent, but as a long-term investment that in normal conditions would appreciate over time. Panama launched an aggressive campaign to attract tourism in the early 2000s, increasing the number of visitors by 67%, even after the economic and travel recession after the events of 9/11. However, the country had been used to the presence of US citizens after a century of canal-related activities, developing a robust road infrastructure and attracting investors with property tax exemptions for new residential development projects over the years, while there are no taxes imposed on offshore companies engaged in business beyond Panamanian jurisdiction.

Also, the real estate market targeted North American baby boomers and retirees who were looking for more attractive options than Miami, Arizona, and New Mexico, the former senior enclaves. In this respect, Panama is granting retiree visas that allow them to remain in the country indefinitely, exemption from property taxes, and access to more affordable medical care, senior discounts on entertainment, transport, prescription medicines, and even mortgages. Nevertheless, retirees comprise only a fraction of the migrant community in Panama (11%), while Central and South Americans account for 65% and Asians around 16%.

Panama has been a long-standing financial hub in the region, with over a hundred banking and financial institutions, a free trade zone, and a skyline dominated by luxury high-rise condominiums, including a marina, hotel, and casino, which are concentrated in four sectors of the city: Punta Pacífica (30%), San Francisco (22%), Balboa Avenue (17%), and Costa del Este (13%).[4] Another factor that explains the real estate market boom is Panama's condition as an offshore tax haven, where banking laws permit tax exemptions and banking secrecy. Anonymous share corporations and shell companies are allowed to own any kind of real estate and financial assets, and they are not required to report sales or transfers, meaning that these transactions remain unaccounted for (Fig. 5.3).

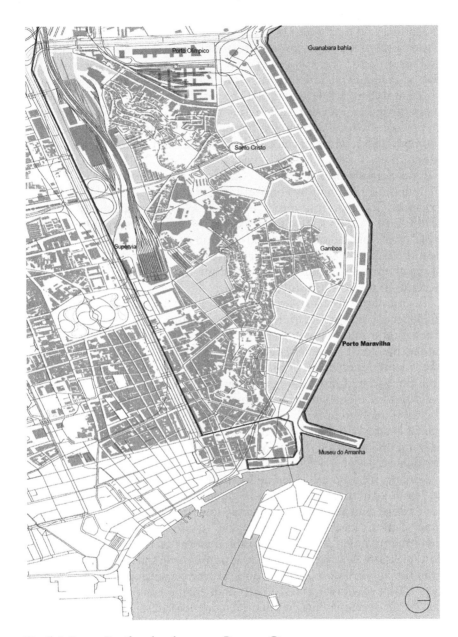

Fig. 5.3 Punta Pacífica development, Panama City.
Source: IDOM-SUMA-CONTRANS, 2018.

On the other hand, the banking sector is duly regulated in areas such as lending and mortgages, holding one of the best ratings in the region. Other attractive provisions are structured finance profits from tax breaks and asset protection, the association of a corporation to every residential or commercial property that the investor buys, avoiding taxation on capital gains while remaining the legal owner of the estate, yet the stock certificates can be transferred to the buyer. Panama's banking sector grants loans to local and foreign borrowers with competitive interest rates (around 8.4%), while developers usually use land as collateral and construction costs are funded with pre-sales (Ganster, 2001, p. 67).

One example of this was the construction of the Corredor del Sur, a 12-mile-long toll highway connecting the financial center with the newly expanded Tocumen International Airport as well as with the eastern part of the city. A Mexican multinational construction corporation (ICA, S.A.) was commissioned to build this infrastructure, and as part of the deal the company was granted rights for the development of 30 hectares of the mainland adjacent to the waterfront, as well as the right to fill in 35 hectares in the contiguous Pacifica Bay.

In the last decades, private real estate developers have heavily influenced land use decisions, lobbying to maximize their investments through the sale of luxury apartments destined for wealthy foreigners rather than local Panamanians. The city's skyline resembles that of places like Miami, Manhattan, or Singapore, which have been built on landfills and along waterfronts, connected by sophisticated highways. High-rise buildings have been instrumental for the real estate market boom, but they also imposed heavy burdens on the city's infrastructure, putting considerable stress on the sewage treatment capacity as well as the provision of water and electricity. They are also creating considerable air pollution problems.

Panama has been heralded as an alternative destination for pensioners and is also capturing the second-home market, for which property values have risen, contributing to displacing low- and middle-income residents from central locations to peripheral areas of the city. Speculative investments have driven the overproduction of units and, as in other cities, it is not even necessary to rent or use those spaces in order to profit from the appreciation of land values, at times resulting in low occupancy rates. Also, new districts have been built that are associated with specific real estate developments (residential, commercial, and/or services), characterized by exclusive homogeneous and enclosed developments. Panama uses the branding and marketing strategy carried on by the private sector (partnered with the government) to promote real estate opportunities to foreign investors through incentives such as the Ley de Condominios program that endorsed densification projects in the central areas of the city without a conclusive urban plan, zoning regulations, or legal framework, in the process increasing gentrification.

It is also worth noting that while high-rise buildings were going up in the central parts of the city, the urban periphery was also expanding rapidly. Scores of uniform housing developments were being built to house the middle-income class, extending the construction boom to the hinterland.[5] The incorporation of middle- and low-income groups was made possible by extending the mortgage financial market to them, creating a valuable asset for the financial markets while responding to a growing demand for homeownership.[6] Therefore, the "spatial fix" that global capital found in the high-rise business district of Panama City also transformed the periphery, serving as a catalyst for a new urban landscape. The residential housing stock grew considerably, mainly as detached homes (54%), apartments (19%), and townhouses (4%), while high-rise buildings accounted for 300 (completed), 124 (under construction), and 191 (projects), mostly concentrated in Punta Pacífica and Punta Paitilla as of 2010.[7]

Panama has a privileged location that connects Central and South America, but also the Caribbean Sea with the Pacific Ocean, attracting international trade as well as becoming a strategic hub for organized crime activities. In this regard, an unrelenting stream of products through the Panama Canal, as well as the existence of a series of isolated islands, has stimulated trafficking of arms, drugs, and people. Also, the central role of Panama in international money laundering networks became evident with the publication of the Panama Papers in 2016, where a local law firm, Mossack-Fonseca, arranged offshore accounts and shell companies for its clients, maintaining the banking secrecy that facilitated money laundering in the country (Lawrence, 2017).

That Panama has become a global landmark with high growth rates in the region there is no doubt. But that this development has been achieved by increasing inequality and through *state capture* by economic elites is also evident. In recent years, Panamanian presidents have praised Dubai, Hong Kong, and Singapore's financial freedom and accelerated economic growth, and yet former premier Ricardo Martinelli (2009–2014) was arrested in Miami for extradition to Panama to face corruption and wiretapping charges, positioning the country in third place on the Transparency International world-rank of corrupt political leaders (Transparency International, 2016).

In what became a major corruption scandal in many countries in the region, Martinelli and his successor, Juan Carlos Varela (2014–2019), were both indicted for accepting campaign donations as well as bribes to grant public contracts to the Brazilian construction company Odebrecht. The two politicians were prominent businessmen in Panama and forged alliances with the private sector that enabled the elites to keep control of the State, even after internal disputes. Freedom House (2020) conceded that Panama has been moving toward democracy, and yet there are still serious concerns about corruption, impunity, money laundering, and tax evasion, especially after the Panama Papers were made public, documenting international fraudulent financial operations by Mossack Fonseca (Obermayer & Obermayer, 2016).

The concentration of wealth among the top elites results in state capture. These groups manage to gain influence through political campaign contributions as well as installing their own people in key government positions and on interlocking company boards, creating a powerful structure to guard their interests (Dal Bó, 2006). These economic elites may influence the State's decisions and orient public policies that are meant to regulate the operation of firms and industries, shape the taxation regime and other things (Innes, 2014). Since countries need private investments to generate growth, establish commercial networks, and create jobs, there is a structural need for collaboration within capitalism. However, there are also mechanisms to increase and maintain political influence through campaign donations, media capture, and particular arrangements, such as the appointment of pro-business officials in strategic positions that define economic and fiscal policies. This complicity is a structural component in the creation of large urban projects, since their operational basis relies on the collaboration of the State and the private sector, for which economic elites take hold of operational and instrumental powers (Fairfield, 2015).

This contributes to widening inequality and favors high-income groups as a "natural" mechanism for concentrating wealth, weakening civil society organizations, and diverting redistributive policies in the making. Private-public partnerships always entail a certain degree of risk since conflict of interests, corruption, and patronage can interfere with the transparency that comes with public office. Hence, the need for transparency mechanisms for the private funding of campaigns as well as restrictions for the appointment of public officers that may have a conflict of interest, one that can discern between public and private aims.

The real estate boom, then, is an expression of more complex economic mechanisms working at different levels, where it is not only the market that explains such phenomena, but their interlocking with the social, economic, and political dimensions. Therefore, the power balance between the State and the economic elite is key to understanding public priorities and procedures, as well as the role that real estate plays in generating value and revenues for the public and the private sector.

A major difference between capitalism and neoliberal capitalism is the role of the State in regulating and enabling the construction of these global devices while also acting as a developer (Valenzuela Aguilera, 2013). Also, Panama illustrates the case of a booming construction sector that at some point may detach from the country's real economy, as has happened in many countries when they are based primarily on finance instead of producing goods and services. As in the case of Asian countries, transnational flows of capital set in motion a local construction industry, whether generating demand or opportunities for investment, creating with this mechanism a self-fulfilling prophecy.

Even though the Panamanian economy relies on trade and related services such as banking, air travel and freight networks, storage, and legal assistance, it is the offshore financial services that serve as the economic engine of the country. The dollarized economy and strong confidentiality laws permit all kind of investors to store their money in this "fiscal haven," including money launderers and corrupt officials from all over the world, who can also invest in real estate through shell companies (Warf, 2002).

Porto Maravilha, Rio de Janeiro

During the 1980s, the American suburban model drove Rio de Janeiro's middle-class developments toward the west, furthering the abandonment and decline of the port district, which increasingly became populated by marginal and homeless residents. The Porto Maravilha project was launched in 2009, just a year before the World Urban Forum 5 took place, with the slogan: "The right to the city: bridging the urban divide." In the following years two international events were planned in the city: the 2014 FIFA World Cup and the 2016 Olympic games, for which Porto Maravilha could serve as a tourist and entertainment hub for visitors, being publicized all around the world. The Porto Maravilha project covered 1,235 acres in the harbor area of downtown Rio de Janeiro as part of a major intervention where infrastructure and public service such as water, sewerage, street lighting, and roads were to be renewed (CDURP, 2020), while the population in the area was expected to increase from the former 30,000 inhabitants to 100,000 in a decade. The project included 44 miles of new streets, 10 miles of cycling lanes, traffic tunnels, cable cars, and a light rail network, aimed at revitalizing the area and gaining prominence among existing Brazilian ports. Most of the land was publicly owned in its various scales (municipal, state, and federal), for which the authorities converged in alliances for the redevelopment project. In many ways it was a classic megaproject (Flyvbjerg, 2014).

The process of appreciation of land values relies on the transformation of obsolete uses and structures into more profitable ones (which may occur even with the sole suggestion of introducing alternative uses), the arrival of higher-income residents through gentrification practices, and the renovation of run-down infrastructure. In Porto Maravilha the revalorization of the area occurred with the installment of high-end shopping and dining facilities, residential and office spaces, as well as cultural facilities, while the informal use of spaces for street vending or similar activities (which may be the only source of revenue for low-income families) was banned and criminalized. Also, the new or renovated public spaces were targeted for middle-class needs and preferences, mixing high culture with exclusive events, while discouraging local street-food vendors, hence marginalizing even further low-income visitors and residents. One of the government strategies was to create an Arts District (Distrito Criativo do Porto), with the hope it would attract the young creative class and get them to set up their ateliers and creative industries. Inspired by Barcelona's Poblenou cultural district, the transformation of the area would certainly bring about the revalorization of land prices and justify public investment in infrastructure.

However, this intervention was condemned as the instrumentalization of culture to legitimize a speculative real estate project that increased disparities and produced expulsions (Souty, 2013). Also, the overall strategy included the renovation or restoration of historic sites and buildings such as the church of Nossa Senhora da Prainha, the remains at the Valongo Wharf, and the hanging gardens. The intervention was rooted in a Eurocentric view, minimizing local – especially African – heritage, and had as its ultimate goal enhancing spatial features that led to appreciating land values, along with the infrastructure for mobility, such as the renovation of the boulevard as well as the construction of a light rail line (Fig. 5.4).

Large urban projects may bring change and opportunities as well as unintended effects such as social segregation and spatial fragmentation. However, the official rhetoric aimed for social integration and diversity, but without putting into effect the necessary mechanisms to achieve such goals. Not only did the increase in land value and rents produce mechanisms of expulsion, but also the transformation of economic activities and the marginalization of informal practices reinforced the conditions for the exodus of local residents. Furthermore, the absence of public policies or zoning provisions to guarantee social housing units left out any last resort to prevent the expulsion of local residents, let alone attracting the low-income population (Prefeitura da Cidade do Rio de Janeiro, 2017).

This was particularly outrageous since most of the Porto Maravilha area was originally publicly owned (75%), and social housing could have been established in those areas at the time of the project design. But it wasn't, resulting in different forms of resistance and criticism for the designers' unilateral and discriminating approach to the local population (Pereira, 2017). Specifically, Afro-Brazilian citizens have expressed their interest in recovering ancient cultural practices such as Capoeira, Carnival *jongo* dancing squads, festivities, and rituals, and, paradoxically, this has resulted in the integration of such manifestations only to attract more tourism (Oliveira et al., 2015).

Porto Maravilha was characterized as an *area of special interest*, overwriting existing urban planning legislation and allowing new parameters in height, uses, and densities. The project was divided into sectors

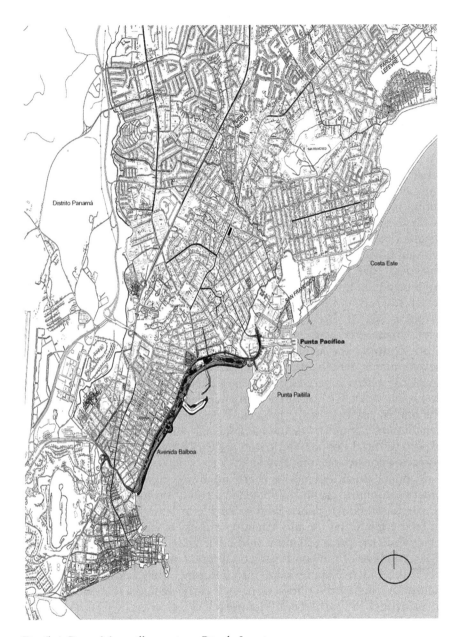

Fig. 5.4 Porto Maravilha project, Rio de Janeiro.
Source: Moro, 2011.

including housing, infrastructure, industry, commerce, entertainment, cultural, and public spaces, organized in two phases: the first funded with public resources and the second carried out through a private entity, the Consortium of Urban Intervention. The project borrowed methods and mechanisms from the private entrepreneurial sector such as management instruments to deal with the negotiations with multiple stakeholders and social actors.

Also, the project demanded dual leadership between public and private partners, whose interests were not always easy to reconcile (Fischer, 1996). To this end, a public-private company was created to manage the project while Concessionária Porto Novo assembled three construction firms to carry out the construction. A special funding scheme was introduced featuring municipally issued bonds (CEPACs),[8] which allowed realtors to obtain additional development rights beyond the limits established by regulations, while financing public works and infrastructure in the area without having to rely on municipal or metropolitan funds. CEPACs were traded as pure financial assets, owned by the semi-public pension and insurance fund known as *Fundo de Garantia do Tempo de Serviço* (FGTS), which holds 8% of formal employers' wages through mandatory deposits (Pereira, 2017). The FGTS purchased all CEPACs and covered the whole managerial costs of the operation, allowing them to have monopolistic control over development rights, as well as the type and volume of developments. Finally, the project was managed by the privately run city agency *Companhia de Desenvolvimento Urbano da Região do Porto do Rio de Janeiro* (CDURP), created with the specific purpose of coordinating the implementation of the project. This agency was created as a mixed-capital public company defined as a special-purpose entity (SPE) with an expected Initial Public Offering on the Stock Exchange and governed by corporate governance precepts.

In major urban operations there are always risks to be taken and, in order to minimize them, FGTS added a condition in which CDURP had to purchase enough public land to enable at least 75% of the CEPACs to be used, offering the fund the opportunity to buy estates at the same cost, while the price of land would be deducted from the CEPAC itself. Also, purchase of the bonds was conditioned on delivering the plots that FGTS considered appropriate for exercising their additional building rights. To conduct the transaction, each party created a real estate investment trust (REIT), the *Porto Maravilha* REIT (owned by FTGS) and the *Região Portuária* REIT (owned by CDURP). Also, FGTS had a long-term business strategy since the fund preferred to partake in the projects and establish partnerships aiming to earn higher revenues from future development projects. Not surprisingly, the fund prioritized top-end buildings and towers since they have higher financial returns, apart from turning real estate into financial assets that provided constant income streams for investors.

Real estate in Porto Maravilha has become one of the safest investments through financialization mechanisms and instruments that enable properties, projects, and rights to become liquid and be traded as such (Aalbers et al., 2017). Also, land can be traded as a financial asset without the intention to use it for development, rent, or use. Among similar funding schemes are Tax Increment Financing (TIF), which monetizes on future tax income from the land appreciation due to the public construction of urban development, amenities, and infrastructure (Weber, 2010). However, this urban land-based financialized speculation may also create undesired territorial outcomes, such as the creation of an empty stock of dwellings that result in major inefficiencies in the real estate market.

Another feature of public-private enterprises is that the role of the State as regulator, mediator, or underwriter of real estate development diverges from its intended objectives and tends to favor certain groups, justifying this action under a framework of exception that operates outside the established regulatory framework (Agamben, 2005). Moreover, these operations require the use of public assets to enhance market dynamism, which somehow prevents the state from undertaking or fulfilling political and democratic obligations.

Urban operations and large urban projects may be considered as part of neoliberal managerial urban governance strategies (Siqueira, 2014), and in the case of Puerto Madero, using a particular financing scheme where CEPACs were acquired by the FGTS pension fund, and different rationalities intersected in the production of urban space.[9] These urban operations depend on flexible zoning rules inside the perimeter establishing floor area ratios (FAR), which are parameters that define construction density in relation to the plot's size, and the basis for the total amount of CEPACs issued. The project defines the amount of bonds to be offered as well as the spatial distribution of development rights within the project area. These bonds are tradable securities that are not attached to any particular intervention, for which they can be traded in the secondary market and monitored by the Securities and Exchange Commission of Brazil (CVM). In other cases, revenues come from the expected increase in property tax revenue, whereas in the Brazilian case, returns from the bonds come from the sale of additional building rights.

In the Porto Maravilha case, flexible regulations and privileged conditions were applied, but with certain conditions. First, revenues from CEPACs had to be reinvested within the urban operation perimeter, which helped improve the living standards in that sector but prevented cross-subsidizing other areas in the city. Later, development rights were traded as securities regardless of which project they were connected to, which allowed greater liquidity but also increased speculation in the financial markets. In the Marxist sense, CEPACs can be considered as fictitious capital since they capitalize on future expectations even if they are issued by the State itself. However, their value depends on market demand while the urban operation lasts, that is, on the increase in land value within the perimeter. It is expected that this reappraisal will increase the demand for additional development rights, capitalizing on future expectations. CEPACs allowed the commodification and financialization of development rights in a market that has heavy State participation, whether by investing in infrastructure, amenities, and services or by intervening in the land market through land acquisition, tax-exemptions, or changes in use and density regulations.

As in most large urban projects of this kind, the project hired prestigious architectural firms in order to attract investors, including Santiago Calatrava for the Tomorrow Museum, Norman Foster for the corporate, Tishman Speyer for the condo towers, as well as other major real estate consultants. All these interventions increased the property values and consequently expelled former residents who could no longer afford the cost of living there. With the State serving as realtor, partner, client, and developer, it was bound to neglect those social claims that may not add value to the value of land and property, and, instead, itself become an active agent of gentrification (Weber, 2010; Fix, 2001). The State may end up overwhelmed by the speculative rationale and favor private capital interests by investing the city's revenues in the project area. The exceptional provisions put in place for these projects may deepen existing patterns of uneven development such as polarization and spatial fragmentation whenever investments are concentrated in certain parts of the city, widening inequality even further. Even if the FGTS pension fund may profit from the project's revenues, it is also true that some of the urban development mechanisms create segregation, gentrification, and polarization, which go against the best interest of citizens, prioritizing shareholders' value over public aspirations (Mosciaro & Santos Pereira, 2017).

A city led by real estate development projects results in the saturation of the urban fabric, the extinction of the remaining public spaces, and the destruction of the built heritage as a costly tradeoff for the idea of modernity. Also, these mechanisms have increased social segregation and spatial fragmentation since the city has not only transformed the skyline but has also created major infrastructure and high-profile centralities, as well as the relocation of lower-income citizens to the periphery of cities. This is how dual cities continue to reproduce, where the capital occupies central locations while social housing developments, irregular settlements, and semi-urban communities occupy the periphery.

The neoliberal rationale is having a major impact on the territory, with the structural reforms that started in the 1980s leading to the creation of land markets that commodified former social land and allowed the concentration of capital in new global hubs (Borras et al., 2012, p. 851), where residential tourism created particular urbanization mechanisms that attracted major investments and fueled the rapid transformations of the urban skyline. In the case of Panama, investors were attracted by fiscal incentives, active real estate markets, banking secrecy, and for some, money laundering (Lawrence, 2017).

The political and economic crisis that the country faced in 2014 affected the real estate markets, including the Porto Maravilha urban plan, which encountered difficulties as the developers and officials sought to commercialize their assets as well as meet their financial obligations. According to various sources, at the end of 2019, 90% of CEPACs were still unsold (CDURP, 2019), despite the rates of occupancy of existing buildings going from 22% in 2017 to 72% in 2019 (O Globo, 2020). This represents a major setback for the project, and is a reminder of the volatility of real estate when it is related to financial markets and, to say the least, is a warning about the limits of projects of this magnitude.

On the other hand, the health crisis represented by COVID-19 rendered even more complicated the use of the office space component of the project, since home office practices are having an impact in the commercialization of corporate space, creating future challenges. Therefore, urban operations are the result of spatial restructuration mechanisms intended to create or reinvent certain parts of the city, but its speculative rationale intensifies real estate market dynamics, which seek increasing profits. Financial schemes look to maximize revenues and include the intervention of the State to guarantee the official endorsement of plans and policies that exempt the project from existing regulations. Once rules, parameters, and restrictions become negotiable, the boundaries between the legal and illegal realms start to fade, and often the public assumes the social costs of misdemeanors. Since the public-private partnerships imply a growing intervention of private agents in the city's management, the emergence of social conflicts out of the private appropriation of collective resources and wealth generated in the city is to be expected, just as has been shown with the Porto Maravilha experience.

Notes

1 The extension is larger than Boston waterfront (41.5ha) and Barcelona (79ha).
2 World-famous Jean-Claude Nicholas Forestier designed Costanera Sur in the early twentieth century.
3 The richest 10% of the population earns 39% of the total national income, while according to the Gini Index, Panama holds the fourth worst income distribution in the region (MEF/BM, 2008).
4 Punta Pacífica is considered the most exclusive residential high-rise sector of the city.
5 It is remarkable that Panama does not have a higher percentage of informal housing (15%), which is uncommon in the rest of the Latin American countries (Angel, 2001). Instead, the urban poor live in public housing developments and substandard housing in central districts of the city.

6 Developers like Grupo Sucasa offered 35–87 square meters in semi-gated developments, increasing social segregation and urban fragmentation in the periphery.
7 High-rises such as the Trump Ocean Club and the Hard Rock Hotel are also known as "cocaine towers" due to their financial links with criminal organizations (Esquirol, 2008).
8 Certificados de Potencial Adicional de Construção (see Sandroni, 2010).
9 In Brazil, urban operations are regulated by the City Statute (Federal Law n. 10.257/2001) and its implementation requires the approval of the local government, which defines its main aspects such as the perimeter of intervention, zoning provisions, and general guidelines.

6 Financial instruments and the city

International flows of capital and urban governance

"In a crisis – said the banker Andrew Mellon way back in the 1920s – assets return to their rightful owners" (Harvey, 2010, p. 11).

In the ongoing neoliberal era, recessions are seen as the product of an unfinished economic liberalization, which demands the deepening of the market rationale, exempting the State from any major responsibility in the face of a crisis. On the other hand, it is argued from Marx's seminal works that privatizing public assets accelerates the circulation of capital, while transferring the control of the national economy to the business sector (Marx & Engels, 2004). Moreover, financial capital drives the deregulation of markets to enable accelerated debt processes, which has led to financial (and later economic) crises, resulting in the repossession of wealth by its "rightful owners." Despite these crises and their devastating effects on the population, capital ends up regenerating and expanding, concentrating its surplus in certain groups and locations, and since capital surpluses have to be re-absorbed in order not to devalue, investment in financial assets once again become imminent.

In the last decades, capital benefited from the integration and deregulation of the banking system as large investment corporations have been assembling pension funds, insurance companies, and investment funds to finance companies, projects, and public infrastructure. In this way, the sources of investment multiply and rescale, driving the design of financial instruments to incorporate new levels of complexity and liquidity, through financial assets such as shares, debt securities, certificates, and variable-yield securities, which enable investors to capitalize future revenues.

But, also, fictitious capital attained a second level through financial derivatives, whose value depends on price variations in the underlying asset, which can be raw materials, minerals, currencies, stocks, stock indexes, rates of interest, etc., and since they can operate through credits, their leverage can multiply either profits or losses. Typically, derivatives serve to hedge future risk through a contract that guarantees the value of an asset against volatility, and yet the market allows speculative practices

DOI: 10.1201/9781003119340-7

such as arbitrage, where profit derives from the price variations in different markets and quotes. These instruments have triggered numerous financial crises, leading billionaires such as Warren Buffett to define them as "financial weapons of mass destruction" (Berkshire Hathaway Inc., 2002) and George Soros to frame them as a "license to kill" (Young et al., 2010), certainly due to their speculative power, which contributed to escalating major financial crises. The over-accumulation of capital requires circulation to avoid devaluation, for which the financial market has been instrumental, up to the point that it equals four times the value of the world GDP, while the market for financial derivatives is equivalent to nine times its value (Prabha et al., 2014). The prominence that financial markets and their stakeholders acquired in recent times has allowed them to steer public policies and define patterns of accumulation in such a way that investments have been channeled to market segments that were not previously considered as the most profitable, and again, without considering social development priorities.

According to Harvey (2004), capital has created particular geographies through "spatial fixes" in the last decades, which allowed investments to be territorialized, locating them in space and transforming the urban environment through buildings and communication infrastructure. Capital surpluses include products, equity, and productive capacity that can be spatially relocated, and to this end, infrastructure plays a central role in the reconfiguration of the territory. Therefore, an extractive rationale applies to various activities, from mineral and energy resources to spaces valued by tourism, either for their natural conditions or their symbolic and cultural value. Infrastructure projects serve as a strategy to slow down the rates of absorption, production, and reproduction of capital, since they involve long-term investments in the territory (Harvey, 2010, p. 26).

Given that infrastructure projects guarantee a privileged link within the market structure, it is expected that economic activities will be attracted by the advantages that their spatial location represents. Communication infrastructure (highways, seaports, and airports) entail an integrating function among production processes, articulating commercial circuits and giving them greater dynamism. It is important to stress that the greatest impacts of connectivity associated with infrastructure are experienced in the territory, both by raising the value of the land and by stimulating the urbanization processes.

The real estate market is deemed the engine of the second circuit of capital accumulation (Lefebvre, 1974), where surplus value is generated not only at the time of its acquisition, but also throughout its useful life, being revalued through the rent, sale, or resale of the property. Being a good that normally requires a mortgage loan, a structural connection is created between the financial and real estate markets, where capital produces a commodity to later become an asset that can be invested as financial capital or as securities in the production of new real estate assets.

Also, the real estate market has a direct relation with the construction of infrastructure and amenities, resulting in different ranges of capital appreciation in the short, medium, and long term. Thus, urban growth can be understood as the expansion of the real estate market, which in turn stimulates the consumption of land and buildings that upholds the consumption of goods and services.

The value of a real estate property originates from the fact that it is a scarce, irreproducible, and geographically located good. That is why the real estate market generates returns through the generation of capital gains derived from the rent of the land or property, its appreciation over time, the infrastructure, and nearby equipment, in addition to the regulatory provisions regarding building densities and land uses. On the other hand, since capital accumulation generates systemic crises (Minsky, 1982), the financial and real estate markets register important imbalances that, from the territorial perspective, generate spatial disparities and massive abandonment in the cases of social housing, shopping centers, apartments, or residential developments.

Real estate crises are preceded by the excess of mortgage credits that trigger property price rises along with the increase in interest rates, producing real estate bubbles that eventually lead the financial market to its limit or to its collapse. The exponential growth in the construction volume and its appreciation in the real estate market are extremely attractive throughout the economic boom, which is leveraged by the financial system. Most of the time, this process involves international large-scale money laundering and corruption networks, along with the complicity of the financial, real estate, and public sectors.

Major real estate crises affect financial institutions (banks, companies, etc.), as well as sectors that are subsidiary to the construction industry, such as developers, construction workers, wholesalers, suppliers, designers, brokers, publicists, etc. Moreover, the crises impact the real economies of the countries, resulting for instance, in a declining GDP, the loss of jobs, and other associated outcomes. As has repeatedly happened, the widespread effect of the market boom reverses during a crisis, which translates into devastating losses and harsh effects on the city. Whenever real estate devaluates, a wave of divestments occurs, since at that point the cost of credit exceeds the value of the property, pushing the owners out of their mortgages.

Territorial dynamics are variegated and when urban policies interact with the financial sector, they can lead the transformation of entire sectors of the city. However, it may be argued that they also incur "creative destruction" of urban areas, displacing the local population whenever capital seeks to bring about more profitable uses within cities. Likewise, real estate products are subject to both their appreciation based on their utility, conservation, aesthetics, or their depreciation due to affectations, deterioration, or disinvestment in the territory. Therefore, developers undertake the conversion of undervalued areas into business districts, large shopping centers, real estate developments, etc.

In Latin America there are cases of regeneration of depressed industrial zones (CDMX, Rio de Janeiro), old train stations (CDMX, Montevideo), deteriorated historic districts (CDMX, Buenos Aires, Quito), waterfronts (CDMX, Santiago, Buenos Aires), or the transformation of garbage dumps (CDMX), territories that due to their location were attractive enough for large-scale real estate investment. These urban redevelopment projects are part of economic restructuring processes in which new productive ecosystems emerge along with different labor markets, whose residents demand their own environments with particular living and consumption spaces.

Real estate markets play a central role in the configuration of cities since they account for the demand for certain types of construction and for specific income segments that convey the production of tailored real estate products. However, this market is subject to both urban policies and existing regulations, which can constrain the urban transformation of some areas (such as historic districts), waterfronts, industrial zones, or promote the massive production of social housing, which before financialization was not considered particularly lucrative for investors.

Along these lines, local governments started to use urban instruments and other mechanisms to drive investments into the real estate market, aiming to enhance the supply of new residential markets to raise the value of certain areas, which allowed value capture through the increase in urban land capital gains (Smolka, 2013). Therefore, investments in infrastructure and amenities enable the reactivation of productive circuits and contribute to the repositioning of historical and cultural assets that are linked to the consumption of both residents and visitors, thereby revaluing the central areas and the return to the built city (Carrión Mena, 1997).

Cities are experiencing a functional transition that is restructuring their productive apparatus, in which the real estate market seeks to provide the spaces required by the new processes of capital accumulation. Due to various circumstances, spatiality has expanded toward teleworking, distance education, and online collaboration, that is, toward the relocation of activities that converge in production or services without this necessarily implying that space is no longer important. The city as a material entity continues to be an engine and fundamental constituent of economic development, and the fact that financial-real estate capital plays such a decisive role for the production and reproduction of the city contradicts the primary function of the State as a guarantor of the common welfare. Large infrastructure projects such as highways, airports, public sector buildings, amenities, and facilities should not exclusively adhere to criteria of profitability and economic efficiency, since they are bound to the contradictions of capital. This is why the government, along with citizens, must assume and resolve these paradoxes on behalf of the population as a whole.

Financialization is playing a major role in the economy of cities, creating new configurations associated with the production, operation, and consumption of the urban built environment, defined as "a pattern of accumulation in which profits accrue primarily through financial channels rather than through trade and commodity production" (Krippner, 2005, pp. 174–175). This economic rationality entails seeking profits and revenues through different mechanisms that are transforming urban dynamics. Instruments such as REITs, public-private partnerships, mortgage-backed securities, and additional building rights are creative mechanisms that allow a certain kind of development.

These instruments can be framed as *sociotechnical devices* able to define specific social relations between the State and citizens (Lascoumes & Galès, 2007, p. 11), which entail a financial rationale, with its own valuation techniques and economic analysis, transforming the power relations among social actors, as well as shaping public policies and the urban realm. Moreover, financial markets provide the means to create and transform the built environment while extracting land rent, and this may only be possible through the endorsement of local and federal governments in regulating, adapting, or removing policy instruments that secure the most profitable form of development through regulatory and fiscal conditions that allow financial markets to unfold (Boyer, 2000). As a result, non-State actors such as financial entities, private equity funds, international investment banks, and rating agencies shape urban and economic policies, and in some cases, large urban projects favor the concentration of investments in specific areas of the city, the expansion of certain companies, as well as the expulsion of former residents through gentrification and relocation mechanisms. Hence, public and private actors will engage in power disputes to readjust the regulatory framework of the city, designing and circulating narratives and representations of such projects, targeting investors' imagination and aspirations, yet these financial strategies remain distant from any comprehensive urban development strategy, due to the lack of effective mechanisms of participation.

Financial instruments for infrastructure

Infrastructure is the material base of the urban economy. This is where productive activities take place, reducing their cost and improving the competitiveness of spatial agglomerations. Yet public investment in this sector may have a differentiated impact on the population, subsidizing capital as well as reproducing power relations and class privileges. Historically, federal, regional, or local governments have funded public works and infrastructure, but in recent decades these efforts have been outsourced to the private sector, which has invested in this sector through financial instruments, which has had territorial, economic, and social implications for the population. This is why it is important to identify the beneficiaries of the financialization processes that are building infrastructure, a question which this section addresses, as it allows us to understand whether productive development is effectively stimulated, concluding that while financial processes favor the reproduction of capital, social benefits also rely on the interaction between the productive base and the commercial circuits articulated with infrastructure.

International development banks and funds consider that in order to close the so-called infrastructure gap in Latin America, it is necessary to secure public and private investments on the order of 5% of GDP (Serebrisky et al., 2015, p. 8; IMF, 2014). In that regard, Standard & Poor's estimates that if national public spending on this sector were to increase by one percentage point, in just three years the size of the economies of Brazil would grow by 2.5%, of Argentina by 1.8%, and Mexico by 1.3% (Christian, 2019).

In order to meet these expectations, the Inter-American Development Bank recommended that the region work toward "strengthening institutional regulatory capacity to develop a portfolio of foresightful projects," as well as "enhancing infrastructure as an asset class to channel private savings to this sector" (Serebrisky et al., 2015, p. 17). Nevertheless, the placement of these assets is never straightforward, since these projects entail high initial construction costs, and unforeseen expenses occur during the construction process, not to mention considerable doubts on the future demand for the services associated with the infrastructure itself. Also, it is important to consider that in Latin American countries debt obligations are usually contracted in foreign currencies, while earnings from the operation or service associated with the infrastructure are in local currency, which among other constraints, prevents developers from finding long-term sources to invest in such assets. In any case, financiers can invest directly in infrastructure projects through existing financial instruments, such as company assets or out of the infrastructure projects themselves.

In this regard, given the long-term investment conditions that infrastructure projects entail, institutional investors have gained prominence in recent years, including pension funds, insurance companies, and investment funds, all of which are financing an important part of the real estate market. The advantages of investing in infrastructure rather than real estate are that the former has less exposure to economic cycles, while also generating stable and recurring returns, which are generally indexed to inflation; and yet private investors challenge the technical capacities of the public sector, as well as its propensity to question toll-rate increases and further investment loans.

From a Marxist perspective, capital generates surplus through over-accumulation that can be absorbed by public infrastructure projects financed by the State in order to further economic growth (Harvey, 2014, p. 154), and yet financial instruments financed by the State itself may also fund these projects. To this end, infrastructure enhances capital, fostering development and innovation, thereby increasing the tax base that allows new investments in physical and social infrastructure. But, as noted before, this process also reproduces inequalities, leading under-served regions into a downward spiral, accentuating their disadvantages. In spite of this, the accumulation of capital requires a particular geography to circulate in space, for which a certain physical infrastructure is necessary to expand and support a qualitative transformation of the territory (Harvey, 2014, p. 157). However, this process may also entail *creative destruction* mechanisms where innovation demands the replacement of previous devices (Schumpeter, 1942), and in which capital is affected by the devaluation of the built environment and induces systematic crises in order to transform it, a process that is viable with the complicity of the State.

According to Serebrisky et al. (2015, p. 7), Latin American countries need to achieve a certain index of sustained economic growth in order to operate effective social development policies and move toward regional integration. To accomplish this, infrastructure that guarantees the provision of services in an efficient and effective way is needed. But since public funding of long-term projects has generally not been sufficient, in recent years subnational governments have explored financing mechanisms such as turning to structured loans, selling of bonds in debt markets using budgetary resources, and using revenues from the operation of projects to meet bank and stock market obligations. These instruments are guaranteed by resource flows that include taxes, rights, uses, and federal transfers, while those generated by projects come from the various fees for electricity, water, energy (PEMEX, CFE), as well as from the operation of airports, dams, highways, thermoelectric plants, or mining.

In recent years, financialization has become a major mechanism for integrating the private sector into infrastructure projects in Latin America, using guaranteed public debt instruments such as stock market certificates or citizen participation certificates, in which debt is backed by these future income streams. A variation of this mechanism is public debt instruments that are not government-backed, but instead the financed project assets guarantee the debt instruments. Other financing alternatives are transfer retainers and federal credit programs, as well as mutual funds and revolving funds, which are financed with preferential credit instruments.

A key element in the financialization process is the growing predominance of agents, institutions, and capital markets in the operation of national and international economies (Epstein, 2005, p. 3), resulting in a rationale that prioritizes finance over productivity, to the extent that the value of financial markets has exceeded GDP value several times. The means of bringing about the financialization of the economy thrived in the eighties and had critical consequences for public finance, since crises occurred on a regular basis. In particular, deregulation allowed a series of fraudulent and predatory practices based on variations of the Ponzi scheme, which used mechanisms unrelated to productive activities, promising investors high rates of return with money taken from later investors. It is important to recall that the deregulation of capital markets brought about a shift of power, from *Fordist production*, where profit was based on mass production, toward the concentration of financial services, the latter becoming the basis for per capita income, to the detriment of productive activities (Harvey, 2007).

According to an Inter-American Development Bank report issued in 2015, infrastructure assets and services constitute the backbone for economic development, competitiveness, and inclusive growth in Latin America and the Caribbean (Serebrisky et al., 2015). A more recent study examined the effectiveness of the existing public- and private-sector financial instruments as a way to invest in infrastructure projects, particularly assessing capital markets in the region (IADB, 2020). However, in the last decade the mechanism most widely used by state governments has been the issuance of stock certificates, allowing the individual participation of their holders through a collective institutional credit by purchasing shares of real estate investment trusts that specialize in infrastructure (streets, public lighting, urbanization, etc.). On the other hand, public-private partnerships are convenient when it is necessary to increase the ratings of bonds or certificates, in cases where particular technical skills are needed, or when the capitalization of operations is required. Generally, projects are backed by financial variations of insurance products that cover insolvency, changes in interest rates, variations in the exchange rates, or depreciation in the value of assets.

In the case of Argentina, a recent Productive Financing Law (PFL)[1] in 2018 aimed to revitalize the capital market and increase institutional spending on project bonds and investment funds. According to the Global Infrastructure Hub created by the G20, which all countries are measured against, Argentina will have an infrastructure gap of USD 358 billion until 2040 (IADB, 2020, p. 20), which might be filled by financial capital. The country's capital market is relatively small compared to other countries in the region and it has a history of high and sudden inflation, not to mention local currency depreciation. All of this has resulted in investors turning to fixed-income instruments.

The PFL has allowed the diversification of products, simplified bureaucratic procedures, and allowed new investors such as micro, small, and medium enterprises (MSME) access to the capital market. Historically, the public sector had driven infrastructure investments but now new instruments such as project bonds (*bonos de proyectos*) allow the purchase of negotiable obligations corresponding to different stages of implementation of the project. Recent regulations permit other financial instruments such as trust funds, investment funds, and other negotiable obligations to operate in the capital market. The law even permitted the use of derivatives to hedge risk, but they are still considered risky instruments compared to sovereign bonds.

In the case of Brazil, debentures are increasingly being used to finance infrastructure because they entail fiscal incentives, although short-term instruments as well as government bonds still dominate the financial market. As in other countries in the region, Brazil is subject to the Basil III regulations, which have limited the ability of private banks to invest in infrastructure projects with its strengthened capital requirements for such projects. Another instrument that is being used in Brazil is the Infrastructure Private Equity Investment Fund (Fundo de Investimento em Participações – Infraestrutura, FIP-IE), which is a private closed-end investment fund where at least 90% of the equity must be invested in infrastructure-related sectors (examples are in the mining industry, thermoelectric plants, etc.). The use of these instruments is part of an effort to compensate for the relative decline in public investments in infrastructure in the last decades.

Chile has financed infrastructure through public investments and mutual funds, endorsed by a strong national public-private partnership framework, as well as by the Comisión para el Mercado Financiero (CMF, 2019), the main regulatory entity for financial markets in Chile, having succeeded in attracting both local institutional investors and international investors. On the other hand, Colombia enacted laws and regulations to enable Valongo Wharf, as well as working to make its infrastructure projects more enticing to institutional investors. Legislation helped to expand the kind of holdings that could act as underlying assets for tradable securities, in the process improving the access to credit, yet the only vehicles for institutional investors to participate directly in infrastructure investments are private equity funds, accounting for 20–30% of infrastructure financing.

In the case of Mexico, subnational governments contracted loans with private or development banks for infrastructure projects that were financially endorsed with federal contributions from the Ministry of Finance and Public Credit up to 100% of their value. However, it was only after the amendments to Constitutional Article 117 that states and municipalities were able to create debt contracts for *productive* investments, although the levels of indebtedness remained in control of the corresponding subnational congresses. To this end, the Federal-Estate Fiscal Arrangements Act established that state governments had to set up an administration trust to receive a percentage of the federal contributions to back up the corresponding loans. Another mechanism frequently used for infrastructure projects has been public-private partnerships, which became widespread in the 1980s for the construction of highways, health care facilities, and energy infrastructure projects. State governments took advantage of the powers granted to them by Constitutional Article 115 for the provision of public services, along with securing resources from the Social Infrastructure Fund, as well as the Municipal Development Fund, among others.

Later, amendments to the Fiscal Arrangement Act in 2013 allowed state governments to back their financial obligations by drawing resources from their federal activities and programmed income to create a trust to place guaranteed loans through a structured payment mechanism. Concurrently, state governments were required to hold good credit ratings in order to maintain their creditworthiness as loan applicants. To this effect, the officials had to provide the rating agency with proof of the evolution of both their income and liabilities, as well as the timely payment of their financial obligations. Positive ratings translated into better credit conditions, including terms and interest rates, besides the fact that the referred structured debt schemes are closely evaluated by the rating agencies, which then allowed them to securitize financial assets such as loan portfolios, accounts receivable, and the flows of future income. The advantages of this scheme are that the securities listed on the stock market acquire greater liquidity, better investment conditions, and allow risk diversification (Fig. 6.1).

Country	Scope of capital market instruments for infrastructure
Argentina	Negotiable obligations (project bonds) Infrastructure investment funds
Brazil	Infrastructure debentures Debentures Investment funds Securitized bonds
Chile	Securitized bonds Corporate bonds Infrastructure investment funds
Colombia	Bonds (corporate or project) Private equity funds
Mexico	Certificates of capital development (CKDs) Energy and infrastructure investment trusts (FIBRA E) Fiduciary security certificates of investment projects (CERPIs) Infrastructure and corporate bonds
Peru	Corporate bonds Private investment funds Real estate investment funds (FIBRA and FIRBI)

Fig. 6.1 Financial instruments in Latin American countries.
Source: IADB, 2020.

Later, new provisions of the pension fund regulatory entity (CONSAR) allowed pension funds and the retirement pension system to invest in energy and infrastructure investment trusts (FIBRA-E) as well as the certificate of investment projects (CERPIs), including the special-purpose acquisition companies aiming to enhance the infrastructure market. AFORES pension funds are allowed to invest 15–20% of their total assets in structured instruments, 10% in investment trusts, and 15–30% in securitized instruments (CONSAR, 2017). Also, the Act for Public-Private Partnerships (2012 and updated in 2018) establishes the conditions and regulations for infrastructure development projects (DOF, 2012). However, capital development certificates (CKDs) are the main vehicle used by AFOREs to invest in infrastructure, holding more than 85% of the CKDs in the market, while CERPIs are issued through a limited offering targeting institutional investors and mostly investing in real estate, energy, infrastructure, and private equity.

To summarize, a major innovation for widening the access to investing in real estate markets are financial instruments, which allow individuals to participate in the profits derived from the rent or sale of real estate without the need to actually own an asset or property. Traditionally, the acquisition of real estate was linked to the possibility of inhabiting the property, but when the objective turns to investment, a return on capital is expected through capital gains over time.

In the first place, it is the investment leverage that allows small investors to access the capital market segment of large real estate projects, where financial entities develop, rent, or sell the properties. In Mexico, this mechanism is operated through financial instruments such as REITs or FIBRAs, CKDs, CERPIs, or infrastructure bonds, which allow significant yields and tax exemptions while handing over the management of the properties to administrators. To put these instruments in context, it is important to understand their legal and financial framework: development banking has always played an important role in this type of investment, yet in recent years private banking made a strong push into the sector, promoting various instruments as well as blending in with pension funds, insurance companies, and investment funds. On the other hand, national development banking institutions are part of the federal public administration, with their own legal features, whose objective is to expand credit while prioritizing key areas for the nation.

International development banks claim that Latin American countries lack sufficient domestic savings, maintain excessive external debt, and possess financial systems that have been ineffective in providing infrastructure, which are some of the features that have characterized peripheral economies against the dominant capital circuits. Starting in the eighties, the *structural adjustments* demanded by the World Bank and the International Monetary Fund produced substantial changes in the structure and operation of financial markets in Latin America. Among those policies, financial deregulation stands out, liberating interest rates and adjusting credit-granting controls, as well as creating a multiple banking system that allowed financing infrastructure projects through direct and structured loans as well as securities.

Financial instruments for infrastructure: Mexico as a case study

There are six institutions that make up the Mexican development banking system: Nacional Financiera (NAFIN), National Bank of Public Works and Services (BANOBRAS), National Foreign Trade Bank (BANCOMEXT), Federal Mortgage Society (SHF), National Savings Bank and Financial Services (BANSEFI), and National Bank of the Army (BANJERCITO). Although these institutions finance most of the infrastructure in the country, in recent decades the Retirement Fund Administrators (AFORES) played a fundamental role as a private financial organization that manages the pensions of workers affiliated with the Instituto Mexicano del Seguro Social (IMSS) and the Instituto de Seguridad y Servicios Sociales de los Trabajadores del Estado (ISSSTE). This pension fund, as an institutional investor, can invest in real estate fiduciary securitization certificates (CFBIs), which nevertheless entails some risk for its affiliates.[2]

These funds, through the Pension Funds Investment Societies (SIEFORES), are the largest institutional investors in the country and validate their assets in structured investment instruments (with a certain degree of risk) and are duly regulated. According to data from the Organization for Economic Cooperation and Development (OECD, 2015) and the National Commission of the Retirement Savings System (CONSAR), Mexican workers receive a pension ranging between 26% and 30% of his or her final salary. Financial specialists insist that to be able to fund their pensions, the fund must invest their capital in projects that will produce higher returns. Among the most used financial instruments are development capital certificates (CKDs), infrastructure and real estate trusts (FIBRAs), and infrastructure bonds, all of which allow financial innovations in financing infrastructure as described next.

Development capital certificates (CKDs)

Development capital certificates (CKDs) were created as an investment vehicle to finance development projects that required intensive capital in the short term with long-term returns, as in the case of infrastructure, mining, and communications projects. These certificates provided investors with considerable flexibility and new alternatives for portfolio diversification, serving as an instrument to mitigate the after-effects of the global financial crisis following the US subprime crisis in the first decade of the millennium. Since its creation in 2009, 81 of these certificates have been placed by their originators for a total value of USD 7,000 billon; top investors were the SIEFORES and to a lesser extent the equity funds, financial groups, hedge funds, sovereign funds, etc.

The SIEFORES manage funds exceeding USD 115 billion, of which 10% – equivalent to USD 11 billion – was allocated both for CKDs funding companies or private investment projects, and investment funds that financed multiple companies or assets. For the latter, the trust administrators had to guarantee that there were potential investors before issuing a certificate. Through these instruments, the SIEFORES were able to access the capital markets, enabling them to generate higher returns for their account holders. However, such investments had to comply with the specifications of the National Retirement Savings System (CONSAR), where each project was required to have a risk assessment by an independent third party at least every quarter before deciding to invest in certificates or any other financial vehicles. Such instruments allowed investors the unique privilege of participating in investment decisions larger than 10% of the total capital of the trust (which is uncommon in private equity funds) by being members of the technical committee and the holders' assembly. In addition, investors expected the administration to invest its own resources along with the issuing trust, with an equivalent percentage in the projects' certificates.

The National Retirement Savings System was created in 1997 to administer workers' retirement and savings funds, while the retirement fund administrators dealt with a diversified portfolio of investments, distributed in government instruments (50%), national private investments (20%), and international equities (16.2%), while the CKDs, FIBRAs, and REITs represent only 6.1% of the total portfolio. The investment in CKDs is open to individuals or legal entities as part of a wide portfolio of clients, and yet those certificates are aimed at financial institutions that handle large capital, such as pension funds, insurance companies, hedge funds, sovereign funds, most major companies, and wealthy financiers. The participation of pension funds in the real estate market allowed them to leverage the projects, since they constitute one of the largest investment funds in Mexico, where by 2020 the financing balance for highways and bridges totaled USD 4,544 billion, while investments in infrastructure via private debt instruments and structured instruments (CKDs, FIBRAs, and CERPIs) accounted for USD 22,104 billion (CONSAR, 2020).

The CKDs concentrate their investments in five sectors: (1) private equity companies (administrators such as WAMEX Capital, Atlas Discovery, Promecap Capital de Desarrollo, AMB México Manager, and EMX Capital I) with 36%; (2) infrastructure such as highways, rail networks, ports, airports, and communications, with 24% (administrators such as Macquaire Mexico Infrastructure, Inmar del Noroeste, Red de Carreteras de Occidente); (3) real estate that includes commercial, industrial, educational, residential, and services assets with 22% (administrators such as Artha Operadora, Vertex Real Estate, Walton Street Equity, Planigrupo Management); (4) renewable and non-renewable energy, hydrocarbons, and electricity 14% (administrators such as Axis Asset Management, Navix de México); and (5) financial assets that include credit and derivatives markets, with 3% (Credit Suisse).

In order to encourage infrastructure projects, the National Infrastructure Fund (FONADIN) participated with up to 20% of the total issuing of CKDs, mainly in the areas of communications, transportation, hydraulic power, environment, and tourism, and also joined in the planning, promotion, construction, conservation, operation, and transfer of infrastructure projects with social impact or economic profitability that correspond to budgeted programs and resources. Investments in CKDs are primarily aimed at private capital projects, which means that they will only distribute returns to the holders when these generate a profit, so those revenues are uncertain and variable and depend on the performance of each project, distributed among private equity companies 28%, infrastructure 24%, real estate 22%, energy 14%, and other 12% (IADB, 2020).

Unlike debt securities, CDKs are not backed by financial liabilities but represent instead a part of the company assets that they finance and, since they are not debt instruments, they are not required to have a credit rating, although they must comply with the standards of both the National Banking and Securities Commission (CNBV, 2014) and the Stock Market Law. These certificates are an alternative way to finance projects that have not yet been developed (also known as *greenfield* or *startups*), usually infrastructure, real estate, mining, business, or technological innovation.

The CKDs are created as long-term instruments (10–50 years), and are compelled to provide the market with sufficient information both on the trust's resources and on the operation of the invested assets. Pension funds have become potential investors in trusts issuing CKDs as long-term financing schemes, even if they entail a certain degree of risk, such as exchange-rate volatility, inflation, unemployment rates, or government actions, as well as internal risks inherent to the projects themselves, such as low-grade planning, increased operating costs, or unforeseen events. In addition, there is no commitment to pay a specific amount of interest or capital, but, instead, returns depend on the profits generated by the financed venture. As the project matures and a state of positive cash flow is reached, investors receive the corresponding profit through the trust in accordance with the provisions of their business plan.

Therefore, the configuration of trusts is essential for understanding the mechanisms through which these instruments operate, since they issue the certificates and pay the expenses associated with the issuance, manage the assets, supervise the investments, and act as intermediaries between shareholders and project managers. In addition, operational transparency is guaranteed with the requirement to deliver quarterly and annual reports on the trust's equity, asset management, and financed projects, in the process disclosing financial information on the investments as they are being made.

Infrastructure and real estate investment trusts (FIBRAs)

Infrastructure and real estate investment trusts (FIBRAs) were created following the model of real estate investment trusts (REITs), which have operated in the United States since the 1960s as conduits for transferring the flows of capital generated by real estate assets and which also offered tax exemptions.[3] The difference between REITs and FIBRAs is that while the former can be constituted as a public and/or private company, the latter have the power to build, acquire, sell, or rent real estate, as well as purchase rights and receive revenues from leasing them. These trusts were launched in Mexico during President Vicente Fox's Quezada administration (2000–2006), aiming to develop the country's infrastructure while securing access for small and medium investors to these instruments.

FIBRAs are defined in the Income Tax Act as "trusts that are committed to the acquisition or construction of real estate with the aim of renting or leasing, or intended to provide the right to receive revenues from such leases, or for granting financing for those purposes" (LISR, 2016, p. 210). This Act regulates the management of trusts, which may specialize in particular segments such as industrial complexes, shopping centers, real estate renting or leasing rights, storage services, educational facilities, offices, executive accommodation, hotels, startups, textile industries, and construction, as well as small and medium-sized enterprises (SMEs).

FIBRAs allow companies to invest in real estate by sharing the property with other investors. It is an innovative instrument that combines a bond and a securities loan. This allows the investor to trade on the stock market as well as receive quarterly revenues (as in the case of a dividend or bond coupon) without having to divert resources from the sale or administration of the property (Office of Investor Education and Advocacy, 2011, p. 1). This way, investors receive regular and consistent dividends and pay fewer taxes, while assets have an upward potential and, above all, have the liquidity to be easily traded, which is a central feature of financialization.[4] Although FIBRAs' performance has fluctuated over time, depending on the particular segment, as a whole they have been highly profitable, since most of these instruments have the mandate to distribute among their investors the equivalent of 95% of returns from the properties in their portfolio at least once a year. On the other hand, the settler (or whoever contributes the real estate) offers their assets through real estate trust certificates (CBFIs), which are placed on the Mexican Stock Exchange market (Grupo BMV, 2012, p. 2).[5]

FIBRAs are exempt from income tax (ISR) and transactions do not pay VAT or real estate transfer tax (ISAI) as long as they have a diverse portfolio and are geographically distributed (Comisión Fiscal, 2017). In addition, investors who contribute their properties to the trust can defer taxes, diversify risk, and maintain their liquidity, since capital gains from trading CBFIs in financial markets are exempt from taxes, under the assumption that an ownership transfer does not take place. Also, a provision exempts FIBRAs from the business flat tax (IETU) during the first years of operation, since the acquisition of real estate is deductible from such taxation.

The appeal of these trusts is that they provide holders with long-term profitability through stable cash compensation as well as capital appreciation through the selective acquisition, construction, and development of a diversified portfolio of well-located properties. In this way, FIBRAs create a portfolio of properties from which it receives a steady income from the rents under the lease contracts. The amount of these revenues will depend on the level of occupation of the leased space, the timeliness of the tenant payments, the time it takes for the vacant spaces to be leased, and the expansion or construction of the properties. In addition, there is the potential appreciation of real estate that will be reflected in the value of the certificate. Another advantage of investing in FIBRAs is that the portfolio approach diversifies their ventures in different markets and terms, in order to reduce fluctuations in the total return of the portfolio, therefore, maximizing returns and minimizing risk (Markowitz, 1952).

Aiming to enhance capital markets for financing public infrastructures, in 2015 the Ministry of Finance and Public Credit (SHCP) announced a new financial instrument for financing infrastructure named *FIBRA E* (infrastructure and real estate trust for the energy sector), issuing fiduciary stock certificates for investment in energy and infrastructure projects. This instrument was modeled after the master limited partnerships in the United States, whose main objective was to securitize mature projects of energy and infrastructure, liberating resources for new projects in the energy sector with it. In its original issuance, different funds such as pension funds (AFORES), institutional and private funds, private banks, and insurance companies created a demand 14% higher than the amount on offer.

In brief, investors acquired rights to a part of the capital of the trust as well as the cash flows derived from it. The trust issued securities to the investors called trust certificates in energy and infrastructure (CBFEs), which have a wide investment spectrum, where petrochemicals stand out as well as the generation, transmission, and distribution of electrical energy; public infrastructure projects such as highways, bridges, railways, seaports, air terminals; water provision, treatment, drainage, and sewerage; and even facilities for public safety as well as prisons and correctional facilities. At the time of creation, various pension funds such as Inbursa, Pensionissste, Profuturo, and XXI-Banorte participated in the issuance of

FIBRA E, which was originally created to finance the construction of the ill-fated New International Airport (NAICM), issuing stock certificates equivalent to 45% of the total amount, of nearly USD 1,500 billion in stock certificates (Alvarez, 2018). These are powerful instruments that aim at strategic sectors of the national economy, allowing the private sector to play a key role in controlling infrastructure previously considered as national security beacons, such as roads, maritime and port terminals, communications, and airports, as well as public security and incarceration facilities. This kind of investments is questioning the previous functions of the State, since prioritizing infrastructure has direct implications on democracy and the sovereignty of nations.

The Inter-American Development Bank has been a key player in identifying possible investment instruments to finance public infrastructure in Latin America, supporting and sometimes directing the region's economies. To this effect, the Bank of Mexico calculates the annual financial statements of the public, private, and foreign sectors in order to track their dynamics over the year, as well as the origin (savings) and destination (investment) of their resources, thus estimating the capacity of the financial system to pay for public infrastructure in the country (BID, 2009, p. 3). The aforementioned report concludes that "Mexico has a diversified financial system that offers a wide range of financial services, which facilitates the structuring of financing schemes for infrastructure projects" (IADB, 2020, p. 4), asserting the potential of the financial sector to fund these projects.

Therefore, real estate operations have found in the securitization of capital an effective mechanism to finance large-scale urban projects, and yet the neoliberal city masks the conditions of reproduction of socio-territorial asymmetries, polarization, and injustices through the commercialization and financialization of space, as a naturalization strategy for real estate speculation in the territory. To this end, new financial instruments, regulatory frameworks, and management models have enhanced the expansion of large real estate ventures under a predominantly economic rationale. There is a territorial impact brought by the financialization of the urban policies that is transforming the configuration of Latin American cities, as the outcome of a particular economic system that is producing important territorial imbalances, where capital accumulation does not return to the processes productive activities that sustain the local economies.

The neoliberal narrative is that Latin American countries are unable to compete in the global economy, favoring their articulation with capital markets, where it is necessary to mobilize resources via modern and efficient infrastructure. Given that in Latin America there are not enough resources for such a commitment, international development banks are proposing the path of debt, opening access to credit, where the future of the country is mortgaged but the circulation of goods in commercial circuits is guaranteed, thus preventing the devaluation of capital surpluses. Consequently, infrastructure may accelerate the circulation of capital through the intervention of financial instruments that will enhance development, but also risk. By financing infrastructure, the State subsidizes the circulation of capital for large companies by deducting transportation costs from the value of the products, and also enhancing the land market by extending the infrastructure to different parts of the city, thus increasing travel and boosting differential rents in the territory.

Financing infrastructure and real estate through financial instruments opens up the possibility of overcoming the temporary conditions of long-term investments, allowing liquidity for these assets, enabling the circulation of capital beyond the barriers inherent to real estate, such as materiality, indivisibility, and geolocation. In addition to facilitating the circulation of capital, infrastructure generates returns as an investment, during the construction process and throughout its entire useful life, generating resources or savings that vindicate the original investment. Financialization comes to solve the limits and contradictions of capitalism (Harvey, 2019), and that is why the action of the State is instrumental to guarantee the presence of the public interest, leaning toward an equal development of the territory, and endorsing instruments for capturing the added value that is generated in the city's financial real estate market.

Financial instruments for shopping centers

Shopping centers are major hubs of the contemporary city, shaping new centralities and places for the distribution and consumption of goods and services. This retail model brought new scales and modes of consumption, as a device for the creation of new enclaves that had a major impact on the structure of cities (Valenzuela Aguilera, 2013). The integration of shopping centers as part of the urban environment entails the adoption of a particular rationality in which global chain stores have become the standard for citizen consumption, positioning names and brands that enhance shopping centers' image and transfer it to virtual commerce.

Retail centers are part of a transnational urban system that is shaping urban space around the world (Sassen, 1991). Since this model replicates consumption patterns, tendencies, and aspirations in most countries, cities adapt and restructure their territory accordingly; citizens tend to standardize their consumption habits, cultural practices, and way of life, while allegedly retaining local traits. In Latin America, shopping centers tend to be more of a leisure and entertainment place rather than solely sites of consumption and, as such, could even lead to new forms of sociability and interaction (López, 2006; Cornejo, 2007; Gasca-Zamora, 2017).

According to the International Council of Shopping Centers (ICSC, 1999), a shopping center is "a group of retail and other commercial establishments that is planned, developed, owned and managed as a single property." They have centralized management and require a common fund to invest in marketing and publicity, offering a variety of goods and services. There are particular regulations in various countries, but shopping center sites occupy no less that 25,000 square meters, usually including a hypermarket, department stores, and several retailers, as well as parking lots that tend to be as large as the interior space. They are classified as *malls*, which are typically enclosed with walkways between facing strips of stores and *open-air strip centers*, which are an attached row of stores or service outlets managed as a single retail entity, align in straight lines, with on-site parking. A more detailed typology of shopping centers according to their scale or coverage is as follows:

- *Neighborhood center*, featuring a supermarket, a department store, and multiplex cinemas.
- *Community center*, with a department store, a large supermarket, multiplex cinemas, a home improvement retailer, and discounted clothing shops.
- *Regional center*, with two or more department stores, a large supermarket, multiplex cinemas, and retail shops.
- *Superregional center*, similar to a regional center, but because of its larger size, a superregional center has more anchors, a wider selection of merchandise, and draws from a larger population base.
- *Fashion/specialty center*, with two or more high-end department stores, multiplex cinemas, restaurants, and other entertainment services.
- *Power centers*, category-dominant anchors, department stores, clubs, warehouse stores, and home improvement retailers.
- *Entertainment/theme centers*, a shopping center with multiplex cinemas, restaurants, entertainment, as well as leisure and sports retailers.
- *Outlet centers*, manufacturers' outlet stores with national and international retailers that sell their products with large discounts.

Malls accelerate and maximize the distribution of goods and integrate leisure and entertainment activities, bring together a multifunctional ensemble of services that enhance the experience of shopping in a continuous, variegated, and complementary flow of consumption. In the case of Mexico and Brazil, shopping malls started in the 1970s as isolated developments, but later in the 1990s they became complementary features of metropolitan extensions. At the beginning, they would compete against other forms of retail established in the traditional parts of the city (like the historic districts), as well as with informal commerce, which has been pervasive all over Latin America (Dávila, 2016).

The expansion of shopping centers in the last two decades can be framed as the result of real estate financialization along with the new urban restructuring that is recycling territories in decline (such as old railway stations, run-down factories, former industrial districts, prisons, and detention centers, etc.), or retrofitting and providing amenities to the expanding metropolitan peripheries in Latin American cities. However, this has also recreated mechanisms of exclusion and social segregation whenever malls are located in exclusive areas surrounded by high-income neighborhoods, becoming fortified consumption enclaves (Caprón & Sabatier, 2007; López, 2006; Salcedo-Hansen, 2003). In most countries in the region, shopping centers are constructed in areas in the city alongside newly built high-income residential developments, areas where their amenities increase land values. This process is also the result of the creation of new centralities and the dislocation of economic activities toward various locations in the city, whether business and corporations, retail, education, industry, tourism, etc.

The first shopping centers in Latin America were born from an association between real estate developers and department stores as a strategy of mutual advantage between two economic groups, and were later classified by their rent per square meter. A next step was the securitization of the projects as well as the introduction of tailored financial instruments such as CKDs and REITs, funded by pension funds, mortgage-backed securities, insurance companies, and corporations such as Black Creek, LaSalle Investments Management, Metlife, and Walton Street, among others (Lizán, 2014), and in the Mexican case, retail-oriented specialized trusts such as Fibra Shop, Fibra Danhos, and Fibra Sendero were created.

According to the ICSC (2015), in the span of five years 326 new shopping malls were constructed in the Latin American region, which have been targeting the upper-middle class. This has had a huge impact in cities with a new form of "retail urbanism" where developments have driven land prices, urban design, and new transit routes (De Simone, 2015, p. 37). A last version of retail centers is located in areas where consumers are low- and middle-income residents who look for hypermarkets and wide-ranging services. In the case of shopping centers, it is estimated that ten of the largest firms control 50% of the market in the US (Retail Traffic, 2010). However, malls serve the State for various purposes, such as formalizing the work force (especially in countries where more than half of the population works in the informal sector) by becoming major employers – FEMSA and Walmart are the top private employers in Latin America – as well as by stimulating the construction sector, with the corresponding permits and taxes.

REITs brought new incentives to mall building since they come with tax exemptions and are able to bid higher for land and properties or concentrate on Class A malls, the top-of-the-line class of retail spaces. These entities usually have a diversified portfolio that allows them to overcome liabilities and losses by transforming existing shopping centers or investing in emerging markets. Also, financial capital is changing the existing rationale, since malls in decline no longer represent the failure of economic mechanisms, but instead can be seen as opportunities to change land uses for higher-paying ones.

Shopping centers stand at the center of a configuration of businesses, service providers, and chains of production that converge at a certain location, attracting further urban development and increasing land values. Despite online shopping and the effects of the COVID-19 pandemic, retail centers remain places where customers are compelled by the experience of shopping along with dining and entertainment. With the massive inflow of capital and imports, shopping mall development has multiplied in Latin American cities in the last decade with a tremendous impact on local retailers and industries. Shopping malls look for the best location, especially near transportation systems and a high density of affluent shoppers, complemented by tax cuts and emergent urban economies, concentrating in large cities, whether following urban expansion or leading it.

According to Baud and Durand (2012), "between 1990 and 2007, the main retailers experienced a tremendous expansion of sales, amounting to an average of 550% growth, whereas the G7 GDP and the world GDP in current US dollars grew by only 112% and 140%, respectively." Thus, even when the retail industry slowed its expansion during that period, it experienced an upward trend in equity that brought increased returns to shareholders, but at the expense of suppliers and workers.

The subprime crisis in the US impacted the Latin American region, yet shopping centers and the retail sector were spared from it, and even countries like Argentina continued to generate revenues in that sector despite the crisis. This apparent contradiction relates to the lack of long-term savings instruments, and in the case of Argentina, the inflation rates led their population to take on a lot of additional consumption credit as credit card loans increased 46.7% between 2011 and 2012 (CEFID-AR, 2012). At the moment, Buenos Aires has 18 shopping centers, of which Cencosud has two, Grupo Village Cinemas S.A. controls two, and the IRSA Corporation owns eight, adopting a dominant position in the sector (Socoloff, 2015). Billionaire George Soros acquired IRSA (Inversiones y Representaciones Sociedad Anónima) some years ago and developed strategic alliances with Goldman Sachs, local pension funds, as well as with Chilean companies, later going public on the New York Stock Exchange and NASDAQ through other branches of the group (Bianchi, 1998). According to that source, the company had increased rents for retailers 30–80%, raising discontent among retailers who claimed monopolistic practices.

According to the National Association of Convenience and Department Stores (ANTAD, 2021), neighborhood centers account for 35% of commercial real estate industry and 70% of shopping centers concentrated in 10 cities in Mexico. Their members hold 78% of total retail sales, leaving to Walmart around 22% of the market. This corporation intends to continue its expansion in Latin America, yet major investments in the retail market were directed toward fashion malls, community centers, and neighborhood centers. Walmart became the first private employer in Mexico and produces 2% of the nation's GDP, yet its business model is based on price reduction, raising productivity, eliminating trade unions, and granting suppliers very low profit margins.

Since in Mexico 80% of commercial transactions are done in cash, it is important to point out that low interest rates, along with banking policies and other strategies to encourage the use of credit cards and promote personal loans, are increasing consumption, as are remittances from transnational workers. Chain retailers such as Walmart, Coppel, and Elektra offer corporate credit cards to buy in their stores, providing access to credit to a group of the population that had been neglected by banks.

In the late 1990s, leading retailers in Europe and North America entered new markets in emerging countries, just before successive financial crises hit Latin America (Coe, 2004; Dawson et al., 2006; Reardon et al., 2007; Wrigley & Lowe, 2007). A decade later, Walmart took a majority stake in Mexican retailer CIFRA in 1997, and its profits increased significantly against its competitors, widening the gap between Walmex and the rest of the retail industry in Mexico, profiting from their trade scale, know-how, and monopolistic capabilities across the country (Durand, 2009). The relative economic stability of recent years has attracted international investment funds with an interest in retail investments, partnering with developers such as Grupo Accion/Planigrupo/Frisa con Kimco Realty Corporation; Grupo Sordo Madaleno/Chelsea Property Group; Calpers/ Hines Interests Limited, and Protego/Discovery (Cushman & Wakefield México).

Shopping malls sell and lease retail space, for which they require large amounts of well-located land. During their lifespan, properties can yield steady flows of income, after which they can be valuable for other kinds of real estate developments, such as residential, health or education facilities, storage, or cultural centers. It is through the cycles of disinvestment and devalorization that financial mechanisms allow major investors to restart a new process of accumulation and concentration of wealth (Smith, 1996). Also, locating shopping malls could also articulate transport networks that enhance economic circuits in the territory, besides appreciating land values and attracting new business and services to the area.

Financial instruments for tourism

Tourism has always been framed as part of a development strategy for emergent economies, leveraging the local workforce and cultural attractions through international loans by the IMF, IADB, and the World Bank.[6] The main axis of this strategy was to use tourism as a way to position these countries' economies at a global scale (Charnock et al., 2014), for which national governments enacted plans and financed programs to promote investment in that sector. This process converted places into objects of tourist consumption, which had an impact on the socioeconomic structure of the site. Even if the goal has been to attract jobs and investments at a local scale, a strategy of differentiation aimed at positioning certain locations within the global circuit of tourist destinations at the expense of similar places that are not included in that selection. Also, tourism became a leading agent in the production and transformation of the territory (Knafou, 2006), especially when framed as an important instrument for capital accumulation (Buades, 2012; Dachary & Arnaiz Burne, 2006), where accumulation by dispossession is prevalent, resulting in land misappropriation, extraction of resources, and socio-ecological conflicts (Harvey, 2004).

Historically, family businesses and developers, along with governmental programs and infrastructure projects, invested in potential tourist sites, then large hotel corporations dominated the market, and later these groups went public in the national and international stock markets, providing liquidity and access to mortgages to operate, and, as a consequence, banking institutions became important shareholders of these groups. At this point, the corporate structures of these groups grew in complexity and channeled part of their revenues to tax havens, so local governments did not even profit from their own tourism industry successes.

In the case of Cancun, Mexico, the federal government took a large loan from the Inter-American Bank of Development (IADB) in the 1960s to create infrastructure (roads, ports, services, and an international airport), as well as investing in hotels, in order to set the conditions for tourism-related development (Arnaiz & Dachary, 2009). In later decades, the intervention of the government also entailed the fortification of the touristic enclaves, the expulsion of local residents, and the devastation of ecosystems. The city of Cancun grew exponentially (and is now close to a 900,000 inhabitants) and with time favored the hotel chains in the appropriation and privatization of the coastline, creating extended, high-profile enclaves. A recurrent element has been the combination of lodging and real estate, with the marketing of timeshare condos, condo hotels, and other second-home residential options.

After Cancun's novelty faded away, new developments and megaprojects shaped the Riviera Maya with a dominant presence of Spanish hotel corporations such as Meliá, Barceló, Riu, and Iberostar, which in the state of Quintana Roo account for 60% of the hotel room stock (Hosteltour, 2009).[7] Tourism also triggered real estate development, and the process of internationalization of hotel corporations brought 30 Spanish real estate groups who invested in México, Panama, Dominican Republic, and Jamaica by 2010 (Blázquez et al., 2011).[8]

In the case of Dominican Republic, the travel industry dates back to the 1970s, when the national government gave generous tax breaks and offered fiscal benefits to investors. In the mid-1980s, a new international airport linked major tourism destinations such as Bávaro and Punta Cana, and all-inclusive resorts were built in these places to attract mass tourism based on economies of scale and global marketing, aiming to convert the island into a world-class tourism (and real estate) hub. Just as in Panama City, the second-home tourist market boomed among retirees as well as other investors in the region, providing tax exemptions, fiscal benefits for businesses, and resident visa facilities. In later years, residential resorts that are integrated ventures and actual urban developments have been attractive to financial capital, resulting in "real estate tourism" (Aledo, 2008), as in the case of Costa Rica (Irazábal, 2018), where this kind of development now has taken up most of its coastline. However, hotel groups do not necessarily own the properties that display their names, while the capital invested may belong to local sources or to international capital. The management is usually handed to operators that implement systems, procedures and standards and employs staff (León-Darder et al., 2011). These operators may be hired by the hotel group or, instead, there could be lease agreements where the revenues go to the operating company that pays a rent to the asset holders in the case of REITs.

During the 1990s, US investors were central to the international tourism market, but from the 2000s on, European groups dominated the industry, especially in Latin America. The economic crises that followed positioned financial capital as a necessary tool for solving problems with debt, enabling them to gain some liquidity by "unlocking" hotel properties. Since costs remained constant and debt kept growing, the hotel assets entered a spiral of devaluation and became true liabilities. Selling the properties to financial groups brought some revenue to hotel corporations, but in most cases these hotels were leased or rented back to the hotel corporations. In this way, financial groups were able to diversify their portfolio of investments, abandoning the already saturated housing market and investing instead in tourist infrastructure (Spolon, 2008; Hidalgo et al., 2016). Their gains were also helped in part because the tourism industry was no longer just providing lodging and services, but revenues started to come from real estate leasing and rents (Navarro et al., 2015).

According to Gotham (2009), this strategy was intended to create liquidity out of "spatial solidity," and with it, investments in real estate at bargain prices due to the urgency of meeting their obligations. With financialization practices, the hotels not only have use value (short-stay lodging), but also actively use their exchange value (the actual price of the building) through the mobilization of property titles, whether by selling them in the primary circuit of capital or by using the real estate investment trusts (REIT) framework on the secondary circuit of capital. Along with private equity funds, REITs accounted for 60% of worldwide hotel transactions (Baker, 2014, p. 2), having short- and long-term consequences on the working environment of the lodging sector. In brief, in the last decades, new mechanisms have emerged among owners and managers, such as lease agreements, management contracts, franchise agreements, and financial instruments such as REITs and private equity funds, leading to a wider strategy of capital circulation. In the near future, the largest international financial entities will own the majority of hotels, becoming a new *rentier* class until a process of devaluation arrives and new financial schemes will be needed for the expansion of capital.

The sector involves international capital and, in the last decades, the financialization of its assets turned it into a key player in the expansion of capital, which supported the local economies and created a new kind of stakeholders that included large hotel and financial corporations, markets, and institutions, becoming one of the biggest industrial complexes (Britton, 1991; Lundberg et al., 1995). According to Euromonitor (2010), the world's hotel industry comprises a large number of independent hotels (85%) competing against large hotel corporations (15%) that nonetheless account for more than half of global sales worldwide. These corporations are dominated by American and European groups, but with an increasing presence of companies originated in China.

Overall, major international crises have had an impact on the tourist industry worldwide, including the subprime crisis, the outbreaks of SARS and Swine Flu, and of course, with COVID-19 in the last few years. Also, tourism has always been associated with transport systems such as airlines, trains, ships, and motor vehicles, at some point creating a vertical integration, which controlled the various stages in the supply chain (Littlejohn, 2003). According to the World Trade Organization (ITC/WTO, 2015), tourism accounted for 6% of total export of goods and services and 30% of commercial services worldwide.

The idea of capitalism is by no means a homogeneous entity with autonomous motivation and program, but its internal configuration needs production as much as it needs consumers. Therefore, the workforce has to be able to buy enough products in order to keep the system flowing (Harvey, 1989b). The case studies examined in this chapter illustrate the fixation of capital in new geographies through particular instruments that create new spaces for investment as part of the second circuit of capital. Tourism as a strategy for development creates wealth but also reproduces inequalities and social polarization, since its main goal is the reproduction of capital and revenues other than social development.

In the case of tourist enclaves, powerful global players such as Barceló, Marriot, and Four Seasons have been increasingly investing in natural retreats and lost paradises in the Latin American region (Honey, 2008). The largest Spanish hotel corporations have been very active in Latin America, among them Barceló, Meliá, and NH Hotels, as well as Hispania, the largest hotel REIT in the Iberian Peninsula. The hotel industry scaled up when financial and private-equity groups such as Blackstone began acquiring major multinational hotel corporations such as Wyndham International for USD 3.2 billion in 2005, Hilton Hotels Corporation for USD 26 billion in 2007, or buying one of the largest hotel real estate investment trusts (REIT) MeriStar Hospitality Corporation, bought in 2006 for USD 2.6 billion (Honey, 2008). An argument can be made that those acquisitions were close to the subprime financial crisis so it is important to understand that the hotel groups were going through a debt crisis in which the only way out was selling their hotel assets, to the point that Hilton at the end did not own the buildings but only operated them.[9] Even when the crisis brought much hardship to the hotel industry, the relative strength of the sector allowed the hotel groups to survive even if they had to sell their real estate assets. Therefore, these groups first used the financial market to leverage their ability to expand, later acquiring the actual assets and corporations.

In the last decades, the hotel industry has diversified its activities beyond short-term lodging, such as hotels and motels, as well as accommodation-related services. The traditional corporate model comprises hotel management, franchise licensing, branding, and marketing, while REITs that specialize in hotels focus on the acquisition, ownership, and operation of real estate. A new actor is platforms like Airbnb, which is an online community marketplace providing short-term rentals that has now contributed more than ten million worldwide bookings through the so-called sharing economy. The impact of this platform on the existing hotel industry is a topic of much discussion, despite Airbnb's claim that 74% of the properties listed on its platform are located outside main hotel districts (Airbnb, 2018). Others believe that a high concentration of these rentals is actually in the primary tourist locations and city centers (Aznar et al., 2016).

Tourism enclaves and resorts have a certain lifespan, after which massive commercial exploitation is no longer viable, losing its appeal and the area starts to decline. At this point, the place has to reinvent itself, reconvert or deteriorate, since the real estate market will look to start new developments with fresh attractions for visitors and investors. The problem with this scheme is that it may no longer be sustainable for the resident population, which has to stay in the place and whose living conditions were not improved while the area was thriving.[10] Meanwhile there were other consequences, including the privatization of the coastline, the increase of segregation and spatial fragmentation, and the destruction of whole ecosystems to build the new developments. Tourism does not convey progress on its own and may only do so as part of a major strategy for territorial development and spatial justice for the region.

Notes

1 Ley de Financiamiento Productivo (N° 27440 v. 11/05/2018), http://servicios. infoleg.gob.ar/infolegInternet/anexos/310000-314999/310084/norma.htm.

2 One possible paradox is that a worker may be financing with his or her pension savings a real estate megaproject that has just expelled him from the neighborhood where he lives.

3 For investors the main advantages of investing funds in REITs are liquidity, security, and return on investment (Standard & Poor's, 2015). Also, REITs allow investors to lease, manage, and operate real estate, in addition to financing development projects or other kinds of assets, offering attractive compensation with a lower risk investment than other instruments such as CDOs or CDSs (NAREIT, 1993).

4 For example, FIBRA Uno had a 100% appreciation of its shares' market price in five years, which added to the quarterly revenue payments, where yields exceeded 140% of the initial investment.

5 There are 17 Investment and Real Estate Trusts currently listed: UNO, MACQUARIE, MONTERREY, DANHOS, HD, NOVA, PLUS; specialized in segments such as HOTEL, INN, SHOP, TERRAFINA, PROLOGUIS, EDUCA, and STORAGE; in addition to FIBRAS E as IDEAL, FVIA, and CFE CAPITAL.

6 According to Hawkins (2009, p. 293) the World Bank participated with 26%, followed by USAID with 16%, the International Development Bank with 12%, and the European Union with 10%.

7 The internationalization of operations among Spanish hotel groups in Central America and the Caribbean was related to the restrictions imposed in Spain (especially in the Balearic area) in relation to the oversupply and obsolescence of the tourist infrastructure (Blázquez et al., 2011).

8 Investments involved hotel buildings, but also buying and selling land, urbanizing areas, and other high-end facilities (golf courses, sports centers, yacht clubs, etc.), in the process creating the conditions for the appreciation of the area, which resulted in capital gains.

9 However, a mix of poor REITs management, overexpansion, and a drop-off in travel after September 11, 2001 were also to blame for the sector's decline (Story, 2005).

10 According to the Human Development Reports of 2005 and 2008, the population of Dominican Republic benefited only marginally from tourism, and even those residents in the most exclusive areas (Bávaro, Punta Cana, and Puerto Plata) saw inequalities and poverty rise (Isa Contreras, 2011; Buades, 2006).

7 Conclusions: sustainable financing practices of cities

Latin American cities have been planned according to real estate market principles, producing contradictory spatial configurations that are continually changing. This is a major public policy issue since one's location in a territory may determine a person's capacity to produce and consume, as it varies in quantity and quality depending on the actual location. For instance, access to goods and services often corresponds to a certain socioeconomic range, which tends to be highly polarized in the Latin American region, which is characterized by major inequalities and spatial fragmentation. In order to find spatial fixes for capital, a number of new urban configurations have been tried; some have proven to be more suitable than others, sometimes dispossessing lower-income residents and introducing capital-intensive activities intended to revitalize and foster more profitable land uses in cities.

The diverse forms of financialization are producing a varied geometry in the configuration of the urban structure, revealing the contradictions of capitalism. However, these configurations remain in constant adjustment, depending on the relocation of capital investments in the city, with certain combinations resulting in territorial disequilibrium. But the urban structure is not shaped by capital investment alone. It is also shaped by the rules and regulations that govern public- and private-sector operations, along with preexisting conditions such as factors of scale, location, and land use organization.

Since the city structure follows market principles and is shaped by a regulatory framework, spatial economic models can explain its underlying composition. However, a financial rationale is driving the various spatial configurations of cities, resulting in the construction of residential buildings, hotels, offices, or commercial centers which are not necessarily driven by the existing demand. Instead, the production of real estate projects responds to financial criteria that seek to maximize returns and activate the flow of capital regardless of broader social concerns. Therefore, when entering the real estate market, financial capital will tend to control its orientation, accelerate its concentration, and centralize its capital gains.

DOI: 10.1201/9781003119340-8

Cities concentrate economic activities and services while creating places for innovation and specialization that enhance the possibilities for economic growth and primacy. Urban structure reflects the economic configuration that lies beneath the surface, but also reveals the existing relations of power, segregation mechanisms, and spatial hegemonies. This is why it is important to assess the urban economic theories that explain the composition of cities throughout different economic periods in history. As economic constructs (by definition), cities require multidimensional explanations that must be evaluated from many angles in order to prescribe the rational and equitable use of resources. They are instrumental for achieving economic growth, but also for accommodating social needs. A better understanding of cities will help the State protect the public interest by enforcing norms and regulations aimed at spatial equilibrium and general welfare, as it must.

One area of recurring concern is economic crises. These have always had a major impact on the spatial configuration of cities, resulting in foreclosures, vacancies, and bankruptcy. It is in the aftermath of crises that capital is transferred and concentrated in certain groups and individuals, deepening the contradictions and inequities intrinsic to capitalism. In order to protect the public interest and minimize the risk of further crises, regulations and controls have to address the operation of real estate markets as well as the financial system that relates to it. The new instruments used to finance urban development and locate capital in real estate markets now play a major role in the configuration of cities at different scales, accelerating processes and creating new urban environments through the investment of capital in the territory and intensifying social disparities and economic polarization.

The financialization of living environments

The housing sector was one of the first segments in the real estate market to integrate financialization mechanisms to boost investments and develop the secondary market. However, the State has traditionally undertaken the provision of social housing in the region, as it is considered a basic human right, first providing the actual units and later establishing a system of subsidies to enhance the demand of the low-income population.

Financial systems in Latin American housing markets are at different stages of development, presenting various sets of options from banking instruments to structured mechanisms; they are considered to have considerable growth potential in the near future. But housing markets are immersed in socially situated economic processes where variations have a direct impact on the living conditions of the most vulnerable citizens. This differs from other financial markets where projections may result in a positive or negative outcome, but will not have a catastrophic impact on the built environment with social, economic, and long-term implications.

The production of space is instrumental for maintaining power and economic relations since it reproduces its historical framework and fixes capital in a territory. Because of this, space can be used as a device to enhance the production of wealth, concentrated in specific areas in the city and appreciating land values while furthering asymmetries in the urban realm. A major contradiction relies on creating affordable housing space and provides access to adequate living environments, which is complicated under the real estate market principles of demand and supply. The problem is that these principles not only respond to actual demand, but also to internal economic dynamics, in which production sometimes exceeds demand and creates a self-reinforcing housing bubble fueled by predatory lending that unleashes the refinancing of mortgages and escalating prices that later result in housing crises.

Along with these procedures most countries in the region have softened financial regulations in order to attract investors, compromising the public interest or any comprehensive city planning concerns in the long run. As real estate in general and housing in particular have entered the global markets, it is even more difficult to disengage their operations from the financial constraints that the articulation with global trade chains entails.

In the case of the leading economies in the region, Brazil and Mexico, the financialization of housing policies produced millions of social housing units (which are generally for sale and not for rent) and yet both countries hold vacant stocks of more than six million units each. Other Latin American countries launched programs that followed the housing model developed in Chile under a public-private scheme of production and later introduced financial instruments to leverage the production of units. Furthermore, the introduction of pension, equity, and insurance capital to finance mortgage securities strengthened the long-term investment capabilities of the system as a whole.

As already mentioned, the value of mortgage-backed securities has relied on the capacity of borrowers to pay back their debt. But due to the volatility of the region's economies, housing crises have been overwhelming for low-income dwellers, who have had to deal with foreclosures and evictions. Since housing has been deemed as a universal right by the United Nations and in many constitutions within the region, any financial rationale that demands strict lending requirements, a clear securitization regime, and foreclosure procedures has been called into question.

In Latin America the production of public housing had been the role of the State, a central means of complying with the right-to-housing mandate. However, their action is now minimized, as much of the low-income population covers this basic need through the informal market, acquiring a piece of land and building a home themselves. Since the 1970s, most of the countries in the region have established housing and development entities to plan and contribute to the affordable housing stock through housing programs and mortgages for the acquisition of social housing units for formal workers with social security coverage.

In the case of Chile, Pinochet's regime introduced a neoliberal economic plan where the State set up a housing market in which the private sector would take a leading role in halving the housing deficit. Even as wages doubled in that country at the turn of this century, inequalities grew deeper and the quality of dwellings remained low. Moreover, housing prices in low-income settlements remained stagnant while those of the middle- and upper-range properties grew exponentially, increasing social segregation and spatial fragmentation. Chile managed to produce 60% of its housing stock just in the last three decades, although it did so by bypassing urban planning constraints, creating dispersed and disconnected housing developments that resulted in adverse territorial diseconomies.

Mexico had previously established a centralized social housing production system, led by national saving funds (FOVI, INFONAVIT, and FOVISSSTE) but during the late 1990s the Chilean privatization approach began to appeal to administrators in the public sector, creating the conditions for the deregulation of the economy. First, the privatization of communal land opened up a way to acquire large reserves of land and later the demand for land and housing was subsidized through mortgages. The production model profited from factors such as scale, specialization, and leveraging, prompting major construction companies to go public on the stock exchange, in the process securing major investments in their ventures. Nevertheless, the very successful housing production system (integrating the various components involved: land acquisition, materials, construction, and marketing) later was a victim of its own success, as relentless operational practices resulted in more than six million vacant houses spread throughout the country.

The financial component of the equation proved to be crucial for the collapse of the social housing market, which started with the subprime crisis in the US and then was exported to Latin America in the following years. In each country the real estate market was subject to different approaches according to its stance toward unregulated markets, global capital, and even toward the previous arrangements between the public and the private sector. In the case of Mexico, the incoming neoliberal administrations in the 2000s fostered the production of housing by real estate developers, providing every means of assistance to these ventures. Yet in the last decade, the administration that followed it restricted the construction of new developments in urban peripheries, in the process accelerating the final collapse of the previous housing production scheme. Since the crisis, Mexico has adopted the Basil III framework standards to hedge risk and promote simple, transparent, and comparable securitization to give investors the ability to evaluate risk as well as possible long-run returns.

Brazil experimented with a broad spectrum of vehicles and instruments to finance real estate, from banking resources to securitized mechanisms, and yet it is still unclear if these procedures served to drive social and economic change across the country. They established a regulatory framework in the late 1990s that allowed the introduction of financial instruments and along with Mexico created a legal framework in which developers and construction companies were able to become publicly traded entities. The exponential growth of the housing market there was fueled by the expansion of mortgage credits for middle-income borrowers as well as the directing of a significant percentage of these loans for second homes, which were used as an investment and not as a residence, in the process financing RMBSs instead of allowing low-income families to own a home.

These construction and real estate companies also profited from the liquidity of global capital markets, and in the case of Brazil the growth of domestic GDP and the decline of interest rates. In this regard, and in order to counteract the upcoming crisis, the *Minha Casa, Minha Vida* (MCMV) program was launched, serving as an economic buffer that enhanced demand while boosting the construction sector and absorbing three-quarters of the subsidies contained in the federal budget. Mortgages in the primary market there were converted to securitization instruments through different certificates and bonds, some carrying restrictions for placing them in the financial markets until they reached maturity.

In the case of Colombia, housing prices have been escalating over the last two decades, while wages have remained stagnant, leaving three-quarters of the housing demand unmet and left to the informal market. Mortgages and subsidies ended up benefiting middle-income housing, which only furthered segregation and inequality, favoring the private production of housing instead of providing decent housing for the 42% of the population that lives below the poverty line.

However, after the Colombian economic crisis of 2004, land and housing prices plummeted 48%, leaving thousands of housing units unsold and 70,000 households facing foreclosure. As we have seen, financial markets produce different and contradictory outcomes: people with mortgages cannot find housing units that match their budget, yet properties stand vacant for years. This contradiction emerges whenever fictitious capital underlies economic transactions, accelerating the collapse of the entire financial structure, all the while being endorsed by governments that usually have vested interests in those ventures. The nature of capital is to seek rent, flowing in large volumes and intensities to the most profitable investments, investments in which the State often retains a major role in regulating, planning, anticipating, and endorsing sustainable development for all.

After the Great Recession, we can agree on the cyclical nature of crises, for which it is necessary to envision different scenarios in order to be prepared for future contingencies. Major investors such as pension funds, insurance companies, and financial entities will continue to capitalize ventures through financial instruments, especially if international banking institutions such as the World Bank, the IADB, etc. keep stressing the need to transform resources into long-term investments. According to these institutions, Latin American countries ought to standardize lending documents and underwriting procedures, foreclosure mechanisms, and the professional criteria for property appraisal to guarantee the quality of the assets backing the securities. However, housing is a constitutional right and remains a major constraint to the market rationale, leaving the State as the responsible entity to represent and look after citizens.

Megaprojects: large investments and the State

In the last three decades, megaprojects have transformed areas of the city in decline into high-tech hubs, innovation and creative districts, important infrastructure elements, or other real estate ventures. Generally, this involves the reconfiguration of space and function, financed by local and international capital. The projects set out to display urban innovation and gain global visibility for the city, which is key to attracting investments as well as positioning the local politicians who drive change. However, megaprojects have also been a vehicle to access global circuits of capital as well as catalysts of growth since they operate as spatial fixes for capital. As with housing developments, these projects have been possible only with the participation of the State, which launches, finances, backs, and promotes the projects along with the private sector. A central narrative legitimizes these ventures on the basis that the country needs a means to access global wealth or even for survival, and this is why public-private partnerships are created that can access capital markets and profit from the revalorization of urban land.

These large interventions have a substantial impact on the territory, transforming its economic, social, and political dynamics, but disregarding the effects that the appreciation of land values brings about (gentrification, segregation, polarization, etc.). They are, in many ways, then, problematic. The projects often lack the mechanisms of participation, as major players make important decisions without consulting the local population. Moreover, the project sites are chosen due to their centrality, which are often areas in decline or needing redevelopment, projects not having a strong social fabric to start with. And the occasion of launching a megaproject calls for exceptional legal and normative frameworks to carry on special interventions that circumvent the existing urban planning instruments. In the end, the projects produce capital gains that are seldom retrieved to benefit the public interest, and instead, the local government builds infrastructure while developers and financial institutions capture the surplus. Finally, land value appreciation furthers differences and inequalities since it produces gentrification and concentration of capital in the territory.

Puerto Madero in Buenos Aires, for instance, has a history of interventions in the waterfront of the city, where periods of progress have been followed by episodes of decay. In the most recent stage of interventions, the area has become a major target for urban redevelopment, in the midst of major economic upheavals and currency devaluation, following successful urban interventions in other parts of the world (Barcelona, Paris, London, etc.). In the case of this project, it has also introduced an environmental edge with the construction of public spaces and the establishment of an ecological reserve in the Costanera Sur. As in other cases, the land revalorization process has expelled low- and middle-income dwellers, while the city government actively participated in the venture, first by selling the land to developers and later by building infrastructure and serving as part of the administrative overseer for the project. Here, as elsewhere, government endorsement has been key to attracting private investments, furthering a speculative logic in which the public sector assumes the costs and most of the risks, while the private sector develops real estate projects aimed at gaining profits that disregard territorial balance, equity, or a better built environment for all.

Santa Fe in Mexico City was first announced as part of an environmental restoration project, in which a new central district would connect the capital city with the global networks that would lead to capital markets, progress, and innovation. As in Puerto Madero, a new governing entity was created to manage development, outline the masterplan, and operate public services, maintenance, and administration. The project went along for four decades with city officials of different political orientations, and the capital accumulation rationale prevailed despite the ideological disparities, probably due to the substantial revenue derived from the annual property tax collection. But these ventures have encountered public resistance whenever the infrastructure or project affects a community nearby, as when the Supervía Poniente highway was designed to pass through low-income settlements and then later modified, or in the case of La Mexicana public park, which was built in exchange for land as well as for a willingness to not build public housing developments near the residential area in the western part of the project. Similarly, the city government was prevented from extending the public transport system in the area (Metro and BRT), probably on the basis that (lower) income mixing would depreciate land values.

Another major example of urban development financed by global capital is Punta Pacífica in Panama City, where financial flows have been constant over a number of years, not least due to the introduction of mechanisms to attract and retain capital through the enabling of Panama as a tax haven, erasing the trail of those resources that may or may not be linked to criminal activities. With a long history of international merchandising (naval, slaves, infrastructure), Panama City has turned into an economic engine in the region, relying on trade, services, and its role as a financial hub. The city has also become an important site for real estate development, mostly targeted to second-home investors, retirees, and investors from neighboring countries as well as from the US, Canada, and Europe.

With a free trade zone and a skyline dominated by luxury high-rise condominiums, the city has been booming with real estate ventures, making Punta Pacífica one of the latest to use financial instruments to increase investments in that sector. The city had a particular history with the US construction and economic exploitation of the Panama Canal, linked as they were for nearly a century before being handed to the Panamanian government, so foreign ventures are not unfamiliar to the country. Not surprisingly, banking there is duly regulated, offering competitive interest rates in mortgages, multiple tax-break opportunities, and flexible urban development requirements.

The recently released evidence of money laundering and tax-evasion practices used by many of Panama's financial system's investors – as well as the imminent increase in inequality and the state capture by the economic elites – casts a shadow over this scenario. Major real estate developments such as Punta Pacífica present a parallel reality to the country's actual situation, where income inequality prevails and the COVID-19 pandemic is having a major impact on the economy. Thus, even when Panama's economy relies on trade and services, tourism and construction, it is the offshore financial services that serve as the backbone of the economy, capturing the fictitious capital that circulates around the world.

Yet another example is Porto Maravilha in Rio de Janeiro, which is similar to Puerto Madero in that both are projects involving the reconversion of a waterfront into major real estate development hubs intended to attract tourism as well as creative industries. The project includes urban interventions such as a light rail network and infrastructure for shopping, dining, and entertainment, while informal activities are banned, leaving the low-income population left aside. The project was legitimized through the instrumentalization of culture by minimizing the African heritage, and furthering segregation mechanisms using diversity as a motif for the urban operation. Just as in the case of Puerto Madero, the local government owned three-quarters of the land to be developed and yet no provision was made to achieve a more balanced spatial distribution, cross-subsidizing vulnerable areas and redistributing costs and benefits.

As in the previous cases, Porto Maravilha created its own governing entity that allowed new parameters of density, uses, and heights, as well as issuing special bonds to finance the project. These instruments (CEPACs) were traded as financial assets and helped to fund the operations that were backed up by a semi-public pension fund. These bonds were conditioned on the execution of building rights and to this effect three special-purpose REITs were created to finance the operations. However, these funds prioritized projects with the highest financial returns, usually condo towers and office and commercial buildings, among other projects, that could be traded as assets in the financial markets. Other financial instruments such as tax increment financing were introduced, allowing capitalized earnings from future taxation of land where development projects were authorized. However, the intervention of private agents in the production of the city brought a rationale other than the residents' perspective, with the imminent emergence of social conflicts associated with the appropriation of the common good.

The instrumentalization of capital in the urban realm

Privatizing public assets accelerates the circulation of capital in which there is a transfer of power from the State to the economic elites that, by using leverage-based techniques for investing, may drive the markets to financial collapse. Among these structured instruments, financial derivatives that bet on the price variations in an underlying asset bring higher levels of speculation to the real estate market. What is always astounding is that after a crisis, financial capital regenerates and expands, usually benefiting some of the major investors and leaving behind a large number of bystanders. The financial system is organized around regulations that change over time, bringing in new levels of complexity, and, most recently, the introduction of algorithms into the transactions known as high-frequency trading.

As Harvey (2004) notes, capital creates particular geographies through *spatial fixes* that have a major impact in the configuration of cities. This is why we analyzed the new geographies of tourist, infrastructure, and commercial functions. Financial capital has been driving the reconfiguration of cities and is being engrained into the economic structure for the production of space. These mechanisms have an impact on the territory since they prompt urbanization processes, socioeconomic dynamics, and land value adjustments, all of which complicate the efficacy of existing urban planning instruments.

Also, and especially in the case of Latin American countries, corruption, money laundering, and other criminal practices continue to alter the operation of the real estate market: first, by impacting land values when massive investments transform certain areas in the city; second, by transgressing the legal framework when planning restrictions are adapted to the needs of capital; and finally, by changing the market rationale when buildings are converted into assets that appreciate in value despite being vacant or occupied. The city undergoes processes of "creative destruction" in which new uses and densities are introduced to revitalize or reconvert an area with some degree of centrality, a process that requires the collaboration or partnership with the public sector. To this end, the financialization of real estate markets has accelerated and re-escalated its operations, providing large investments in short periods of time and creating brand-new sectors in the city with a cosmopolitan allure, that nevertheless remain isolated from the rest of the city.

Infrastructure provides the opportunity for long-term projects, which is the kind of low-risk investment that has less exposure to economic cycles while generating stable and recurring returns. These projects are meant to go along with broader development goals since they reduce distribution and exchange costs, increase competitiveness, and are key for increasing the regional GDP. Also, given the nature of long-term ventures, insti- tutional investors have gained prominence in recent years, including pension funds, insurance companies, and equity funds, as has the use of innovative financial instruments in Latin America with different degrees of complexity. In this regard, Mexico has been a forerunner in the cre- ation of development capital certificates (CKDs) as well as real estate investment trusts (FIBRAs), already used to finance highways, airports, and other major infrastructure projects. In the neoliberal narrative, these projects enable emergent economies to compete in the global market, giving them the ability to mobilize products and resources across the terri- tory, while minimizing transportation costs. Moreover, at the other end of the chain of production stand the points of distribution that integrate con- sumption, leisure, and entertainment in shopping centers. These enclaves have attracted new centralities and impacted land value by recycling old infrastructure (derelict factories, train stations, etc.) or transforming the what already existed. In various cases, tailored financial instruments were based on REITs and funded by pension funds, insurance companies, and corporations such as Black Creek, LaSalle Investments Management, Metlife, and Walton Street.

Latin America has had a major boom in the construction of new shopping centers in recent years that include hypermarkets and a wide range of services that appeal to various socioeconomic ranges. The latest trend of fostering consumption credit has resulted in increasing use of credit card loans that, along with transnational remittances, are being used to buy goods and services. However, it is the cycles of disinvestment and devalorization that allow the reconfiguration of the territory, in a con- tinuous process of re-appropriation of wealth.

One other area of importance is tourism, which has long been a major economic engine in Latin American countries, as it attracts international capital and generates complementary activities for the local economy. At best, this strategy positions tourist hubs at a global scale, where governments are usually eager to support the construction of major infra- structure, providing the conditions for investment and the legal means for undertaking major ventures. As an important instrument for capital accumulation, tourism favors certain places, creating tension with similar venues, and has an impact on land values in most of the region where land becomes a valuable asset that may generate socio-ecological conflicts.

The travel industry started out in the region as a family business. But it was then taken over by international hotel corporations and then, more recently, by financial institutions, growing in complexity and creating an intricate system of tax avoidance (that is to say, a way of legally reducing your taxable income). These corporate structures are not contributing to the local tax base but instead outsourcing their profits to tax havens, creating a shadow economy in which the exploitation of natural or cultural resources does not create wealth for the local population.

In cases like Cancun (Mexico), the federal government created infrastructure (roads, services, and an international airport) with loans from the IADB as part of a plan to build a major city-resort that would attract international investment, wealthy migrants, and top hotel chains. Spanish real estate corporations were the first to consider the area a high-profile enclave and they took hold of the market, for many decades dominating the Latin American scene. However, after the subprime crisis had affected not only the US but also southern Europe, hotel corporations accumulated so much debt that they ended up selling their properties, and later renting them back, all the while keeping the operation and management of the hotel brand. Instead, financial corporations captured revenues from leasing, renting, and the property value itself. These corporations used REITs to join the secondary circuit of capital, accounting for 60% of worldwide hotel transactions (Baker, 2014, p. 2), creating new stakeholders that joined the financial tourist complex, which then became dominated by North American, European, and, lately, Chinese companies.

Those in the tourist industry have always relied on transport systems and infrastructure, with a partnership with corresponding governments as the key to their success. Yet their main goal has been the reproduction of capital and revenues, leaving aside social development concerns. This is not unlike the real estate rationale where firms seek to maximize profit that is used by major financial firms such as Blackstone and other leading global businesses who leverage their investment capital on the assembled capital of pension funds, large institutions, and individuals, and that in the last decades have been part of major transactions in a wide range of ventures. As in other real estate domains that we have examined, major tourist enclaves are in the process of reinventing themselves, looking to attract capital ventures and in many cases migrate to new and more appealing destinations. Therefore, it is clear that tourism does not convey progress on its own and may only do so as part of a major strategy for territorial development and spatial justice for the people.

In conclusion, capitalism is transforming the operation of the financial sector, prioritizing the credit market that on the one hand expands liquidity, and on the other favors speculation activities that allowed the appropriation of financial gains. The opening of deregulated markets, trade agreements, and other economic and regional integration shaped Latin American financial systems to the dominant economies. International development banks promoted the financialization of real estate markets as a capitalist strategy aimed at overcoming the tendency toward overaccumulation while catering to powerful capitalist interests. In this sense, the neoliberal period introduced a *financial turn* to market-led solutions, heavily supported through State funding and regulatory framings, to solve the contradictions that emerged from the broader dynamics of capital accumulation (Harvey 1999). As a result, capital interests in most countries in the region continue to prosper with the unconditional support of the State, through loans, guarantees, regulations, and tailored policies.

The right to the city is a prospect that has been pushed by social organizations around the world but that has also been hindered by the financial dynamics of capital. These dynamics have become a driving force as major global investors, international development banks, and Wall Street traders are investing heavily in urban projects, and as the World Bank, through its various subdivisions, has steadily promoted this financial market complex. This configuration has resulted in the overproduction of housing, unleashed segregation and gentrification, and deepened the existing socioeconomic divides in the city, which may fit with a rationale of maximizing profit but stands far from moving us closer toward a just and inclusive city that would provide a livable environment for all.

Global capitals have encountered a compelling and lucrative target for investments in Latin American real estate markets, where the building stock acts as a reserve of value for financial assets thanks to local governmental support, the globalization of economic processes, and the digitalization of financial procedures. In this context, the revitalization of neighborhoods, the creation of new touristic enclaves or the refunctionalization of districts produce the appreciation of land prices that enhance real estate as a prime asset in the financial markets. Global financial corporations favor long cycles of capital circulation that real estate and major infrastructure can provide, especially when they are endorsed by local governments. It is important to mention that tax havens assemble financial capital flows from developed countries as much as from developing countries, as was evidenced by the Panama and Pandora Papers in recent years, which also find the real estate market as one of the most profitable ventures in which to invest.

This is how financial capital finds multiple mechanisms to access financial rents in economic structures that only partially partake of this rationale, but still manage to adapt to the circumstances and conditions operating in each of the Latin American countries. Also, international banking institutions such as the World Bank have been eager to instrumentalize the mechanisms to foster global capital investments, such as funding the creation of real estate investment funds, mortgage-backed securities, etc.

Also, the State played a crucial role in setting up the financial framework in which real estate transactions can thrive and attract global capital for investments, which include flexible regulations, loans and mortgages, grants and subsidies, as well as building infrastructure and extending basic services to the areas of intervention. Paradoxically, the availability of financial capital and credit increases land values, generating a self-fulfilling prophecy of ever-growing real estate appreciation, disregarding the actual demand. This appreciation in value leaves most of the population behind, producing gentrification, expulsions, and displacement, leaving centralities for those who can pay to live in them.

Financial capital has grown exponentially through digital flows and this has been a global phenomenon that is taking real estate as a vehicle to reproduce fictitious capital. This has important implications for cities' governance since public policies are directed toward citizen's welfare while finance aims to create value at any cost. Therefore, different rationalities are in place responding to divergent interests, and only through political commitment can regulations be put in place to guarantee a just, responsible, and inclusive city.

References

Aalbers, M. B. (2009). The sociology and geography of mortgage markets: Reflections on the financial crisis. *International Journal of Urban and Regional Research*, 33, 281–290. Available at https://doi.org/10.1111/j.1468-2427.2009.00875.x, accessed on June 7, 2017.

Aalbers, M. (2016). *Financialization of housing, a political economy approach*. London: Routledge.

Aalbers, M. B. (2017). The variegated financialization of housing. *International Journal of Urban & Regional Research*, 41 542–554.

Aalbers, M. B. (2019a). *Financial geographies of real estate and the city. A literature review*. Financial Geography Working Paper Series #21, FinGeo, University of Leuven.

Aalbers, M. B. (2019b). Financial geography II: Financial geographies of housing and real estate. *Progress in Human Geography*, 43(2), 376–387.

Aalbers, M. B., Van Loon, J., & Fernandez, R. (2017). The financialization of a social housing provider. *International Journal of Urban & Regional Research*, 41(4), 572–587.

Abramo, P. (1997). *Marché et ordre urbain: Du chaos à la théorie de la localization résidentielle*. Paris: L'Harmattan.

Agamben, G. (2005). *State of exception*. Chicago: University of Chicago Press.

AIG SunAmerica Life Assurance Co. (2002). Form 10-K, for the fiscal year ended December 31, 2002. Available at https://last10k.com/sec-filings/6342, accessed on May 18, 2019.

Airbnb (2018). *About us – Airbnb newsroom*. Available at https://press.atairbnb.com/about-us/, accessed on July 10, 2021.

Aledo, A. (2008). De la tierra al suelo: La transformación del paisaje y el nuevo turismo residencial. *ARBOR Ciencia, Pensamiento y Cultura*, 184(729), 99–113.

Alonso, W. (1964). *Location and land use: Toward a general theory of land rent*. Cambridge, MA: Harvard University Press.

Alvarez, S. (2018). Cómo te afecta que las afore inviertan en el nuevo aeropuerto. *Expansión*. Available at https://expansion.mx/dinero/2018/04/02/como-te-afecta-que-las-afore-inviertan-en-el-nuevo-aeropuerto, accessed on May 30, 2018.

Andrews, E. (2008, October 24). Greenspan concedes error on regulation. *New York Times*.

Angel, S. (2001). The housing policy assessment and its application to Panama. *Journal of Housing Economics*, 10(2), 176–209.

ANTAD (Asociación Nacional de Tiendas de Autoservicio y Departamentales) (2021). Available at https://antad.net/indicadores/indicantad/, accessed on June 3, 2021.

AMB (Asociación Mexicana de Bancos) (2020). *Reporte de Estabilidad Financiera, Diciembre.* Available at https://www.banxico.org.mx/apps/ref/2020/g-nivel-de-morosidad-de-la-.html, accessed on February 15, 2022.

Arbeláez, M. A., Camacho, C., & Fajardo, J. (2011). *Low income housing finance in Colombia.* IDB working paper series 256. Washington: Inter-American Development Bank.

Arnaiz, S. M., & Dachary, A. C. (2009). *Geopolítica, recursos naturales y turismo. Una historia del Caribe mexicano.* Guadalajara: Universidad de Guadalajara.

Artle, R. (1972). Urbanization and economic growth in Venezuela. *Papers and Proceedings of the Regional Science Association,* vol. 27, pp. 63–93.

Ashton, P. (2009). An appetite for yield: The anatomy of the subprime mortgage crisis. *Environment & Planning A,* 41, 1420–1441.

Aznar, J., Sayeras, J. M., Rocafort, A., & Galiana, J. (2017). The irruption of Airbnb and its effects on hotels' profitability: An analysis of Barcelona's hotel sector. *Intangible Capital,* 13(1), 147.

Baker, T. (2014, April 2). REITs hovering but wary of European landscape. Available at http://www.hotelnewsnow.com/Article/13441/REITs-hovering-but-wary-of-Europe-landscape, accessed on February 12, 2020.

Balchin, P. N., Bull, G. H., & Hieve, J. L. (1995). *Urban land economics and public policy.* New York: Palgrave.

BACEN (Banco Central do Brasil) (2022). *Index of economic activity of the Central Bank.* Available at https://www.bcb.gov.br/en, accessed on February 15, 2022.

Banco de la República (2009). *Reporte de estabilidad financiera – Septiembre 2009,* Bogotá: Banco de la República.

Banco de la República (2016). *Reporte de estabilidad financiera – Septiembre 2016,* Bogotá: Banco de la República.

Banxico (2011–2016). *Boletines estadísticos de Banca Múltiple.* Available at http://portafoliodeinformacion.cnbv.gob.mx/bm1/Paginas/boletines.aspx, accessed on February 15, 2022.

Basel Committee on Banking Supervision (2016). *International convergence of capital measurement and capital standards. A revised framework,* comprehensive version. Basel: Bank for International Settlements.

Baud, C., & Durand, C. (2012). Financialization, globalization and the making of profits by leading retailers. *Socio-Economic Review,* 10(2), 241–266. Available at https://doi.org/10.1093/ser/mwr016, accessed on April 22, 2019.

BBVA Research (2016). *Situación inmobiliaria México 1S16.* CDMX: BBVA. Available at https://www.bbvaresearch.com/wp-content/uploads/2016/04/1604_SitInmobiliariaMexico_1S16.pdf, accessed on February 15, 2022.

Benchimol, J. (1992). *Pereira Passos: Um Haussmann tropical. A renovação urbana da cidade do Rio de Janeiro no início do século XX.* Rio de Janeiro: PCRJ/SM.

Berkshire Hathaway Inc. (2002). *2002 Annual Report.* Available at http://www.berkshirehathaway.com/2002ar/2002ar.pdf, accessed on July 25, 2017.

Bertaud, A. (2015). *The spatial distribution of land prices and densities: The models developed by economists.* Working Paper #23, Marron Institute of Urban Management, New York University.

Bianchi, A. (1998, February 26). Soros sube los precios en shoppings. *Diario La Nación*. Available at https://www.lanacion.com.ar/economia/soros-sube-los-precios-en-shoppings-nid88935/, accessed on February 10, 2019.

BID (Banco Interamericano de Desarrollo) (2009). *Panorama del financiamiento de infraestructura en México con capitales privados*. Washington, DC: BID.

Blázquez, M., Cañada, E., & Murray, I. (2011). Búnker playa-sol. Conflictos derivados de la construcción de enclaves de capital transnacional turístico español en El Caribe y Centroamérica. *Scripta Nova: Revista electrónica de geografía y ciencias sociales* 15. Available at https://raco.cat/index.php/ScriptaNova/article/view/244215, accessed on July 24, 2021.

Boletín Oficial de la Ciudad de Buenos Aires (O:M.44.945, 1991). *Ordenanza Área de Protección Patrimonial Antiguo Puerto Madero Art. 12, sancionada por el Honorable Concejo Deliberante*. Available https://boletinoficialpdf.buenosaires.gob.ar/util/imagen.php?idn=87223&idf=1, accessed on February 15, 2022.

Borja, J. (2001). El proyecto metropolitano: El manejo de una variable geométrica. In M. Freire & R. Stren (Eds.), *Los retos del gobierno urbano* (pp. 18–25). Bogotá: Banco Mundial/Alfaomega.

Borja, J., & Castells, M. (1997a). *Local y global. La gestión de las ciudades en la era de la información*. Madrid: Taurus.

Borja, J., & Castells, M. (1997b). *The management of cities in the information age*. London: Earthscan/United Nations Centre for Human Settlements.

Borras, S., Franco, J., Gomez, S., Kay, C., & Spoor, M. (2012). Land grabbing in Latin America and the Caribbean. *Journal of Peasant Studies*, 39(3–4), 854–872.

Bourdieu, P. (1984). *Distinction: A social critique of the judgement of taste*. Cambridge, MA: Harvard University Press.

Bourdieu, P. (2003). *Las estructuras sociales de la Economía*. Barcelona: Anagrama.

Boyer, R. (2000). Is a finance-led growth regime a viable alternative to Fordism? A preliminary analysis. *Economy & Society*, 29(1), 111–145.

Braga, J. C. S. (1997). Financeirização global. O padrão sistémico de riqueza do capitalismo contemporâneo. In M. C. Tavares & J. L. Fiori, *Poder e dinheiro. Uma economia política da globalização* (pp. 195–242). Petrópolis, Brasil: Editora Vozes.

Braudel, F. (1982). *Civilization and capitalism, 15th–18th century: The wheels of commerce* (Sian Reynolds, Trans.). Berkeley: University of California Press.

Brenner, N. (2009). Restructuring, rescaling, and the urban question. *Critical Planning*, 16, 61–79.

Britton, S. (1991). Tourism, capital and place: Towards a critical geography of tourism. *Environment & Planning D: Society & Space*, 9, 451–478.

Buades, J. (2006). *Exportando paraísos: La colonización turística del planeta*. Palma de Mallorca, España: La Lucerna.

Buades, J. (2012). Clima & Mediterráneo & Turismo en el siglo 21. In E. Navarro & Y. Romero (Eds.), *Otras miradas, otros turismos*. Málaga: Universidad de Málaga.

Buffet, W. (2003, February 21). *Letter from Warren Buffett, Chairman of the Bd., Berkshire Hathaway Inc., to the Shareholders of Berkshire Hathaway Inc.* p. 15.

Burgess, E. W. (1967) [1925]. The growth of city: An introduction to a research project. In Robert E. Park, Ernest W. Burgess, & Roderick D. Mckenzie, *The city* (pp 142–155). Chicago & London: University of Chicago Press.

Camagni, R. (2004). *Economía urbana*. Barcelona: Antoni Bosch.

CAPMSA (Corporación Antiguo Puerto Madero S.A.) (1992). *Anteproyecto urbano para Puerto Madero. Memoria histórica y urbanística*. Buenos Aires: CAPMSA.

CAPMSA (Corporación Antiguo Puerto Madero S.A.) (1999). *Santiago de Chile: Un modelo de gestión urbana 1989–1999*. Buenos Aires: Ediciones Lariviere.

Capel, H. (2002). *La morfología de las ciudades. I Sociedad, cultura y paisaje urbano*. Barcelona: Ediciones del Serbal.

Caprón, G., & Sabatier, B. (2007). Identidades urbanas y culturas públicas en la globalización. Centros comerciales paisajísticos en Río de Janeiro y México. *Alteridades*, 17(33), 87–97. Available at http://www.redalyc.org/articulo.oa?id=74712772008, accessed on September 30, 2018.

Cardoso, A. L. (Ed.) (2013). *O programa Minha Casa Minha Vida e seus efeitos territoriais*. Rio de Janeiro: Letra Capital.

Cardoso, E. et al. (1987). *História dos bairros: Saúde, Gamboa, Santo Cristo*. Rio de Janeiro: Index.

Carrión Mena, F. (1997). *El regreso a la ciudad construida*. Lima: DESCO.

Caskey, J. P., Ruiz Duran, C., & Solo, T. V. (2006). *The urban unbanked in Mexico and the United States*. Policy Research Working Paper 3835. Washington, DC: World Bank.

Castells, M. (1974). *La cuestión urbana*. Madrid: Siglo Veintiuno.

CDURP (Companhia de Desenvolvimento Urbano da Região do Porto do Rio de Janeiro) (2019). *Relatório trimestral de atividades*. Available at http://www.portomaravilha.com.br/conteudo/relatorios/2019/outnovdez.pdf?_t=158807270, accessed on August 3, 2021.

CDURP (Companhia de Desenvolvimento Urbano da Região do Porto do Rio de Janeiro) (2020). *Porto Maravilha*. Available at http://www.portomaravilha.com.br/portomaravilha, accessed on August 1, 2021.

CEFID-AR (2012). Informe mensual de préstamos al sector privado no financiero (SPNF) y otros indicadores económicos y monetarios. *CEFID-AR*, 9(104). Available at http://www.cefid-ar.org.ar/documentos/CEFID-AR-Informe_104.pdf, accessed on June 30, 2021.

Cepni, O., Dul, W., Gupta, R., & Wohar, M. (2021). The dynamics of U.S. REITs returns to uncertainty shocks: A proxy SVAR approach. *Research in International Business and Finance*, 58, Article 101433. 10.1016/j.ribaf.2021.101433.

Cerutti, E., Dagher, J., & Dell'Ariccia, G. (2015). *Housing finance and real-estate booms: A cross-country perspective*. Washington, DC: International Monetary Fund.

Charnock, G., Purcell, T., & Ribera-Fumaz, R. (2014). *The limits to capital in Spain: Crisis and revolt in the European South*. New York: Palgrave Macmillan.

Cheikhrouhou, H., Gwinner, W. B., Pollner, J., Salinas, E., Sirtaine, S., & Vittas, D. (2007). *Structured finance in Latin America. Channeling pension funds to housing, infrastructure, and small businesses*. Washington, DC: The International Bank for Reconstruction and Development/The World Bank.

Christaller, W. (1966) [1933]. *Central places in Southern Germany* (C. B. Baskin, Trans.). Englewood Cliffs, NJ: Prentice-Hall.

Christian, C. (2019). *The guide to infrastructure and energy investment*. London: Law Business Research.

Ciccolella, P. (1999). Globalización y dualización en la Región Metropolitana de Buenos Aires. Grandes inversiones y reestructuración socioterritorial en los años noventa. *Revista EURE*, 25(77), 5–27.

Clavijo, S., Janna, M., & Muñoz, S. (2004). La vivienda en Colombia: Sus determinantes socioeconómicos y financieros. *Borradores de Economía* No. 300, Banco de la República. Available at www.banrep.gov.co/docum/ftp/borra300.pdf, accessed on January 18, 2016.

CMF (Comisión para el Mercado Financiero) (2019). *Qué es la CMF*. Available at http://www.cmfchile.cl/portal/principal/605/w3-article-23900.html, accessed on October 10, 2018.

CNBV (2014). *Reglamento interior de la Comisión Nacional Bancaria y de Valores* (Diario Oficial de la Federación el 12 de noviembre de 2014). Available at https://www.cnbv.gob.mx/Normatividad/Reglamento%20Interior%20de%20la%20Comisi%C3%B3n%20Nacional%20Bancaria%20y%20de%20Valores.pdf, accessed on May 23, 2018.

Coe, N. (2004). The internationalization/globalization of retailing: Towards an economic-geographical research agenda. *Environment & Planning A, 36*, 1571–1594.

Comisión Fiscal (2017). *Ley del impuesto sobre la renta 2017. Texto y comentarios*. México: Instituto Mexicano de Contadores Públicos.

CONSAR (2017). *Límites del régimen de inversión*. Available at http://www.consar.gob.mx/gobmx/Aplicativo/Limites_Inversion/, accessed on May 21, 2019.

CONSAR (2020). *El sistema de ahorro para el retiro al cierre de 2020*. Comisión Nacional del Sistema de Ahorro para el Retiro. Available at https://www.gob.mx/consar/es/articulos/el-sistema-de-ahorro-para-el-retiro-al-cierre-de-2020-261475?idiom=es, accessed on July 27, 2020.

Cornejo, I. (2007). *El lugar de los encuentros: Comunicación y cultura en un centro comercial*. Mexico City: Universidad Iberoamericana.

Cubeddu, L., Tovar, C. E., & Tsounta, E. (2012). *Latin America: Vulnerabilities under construction?* IMF Working Papers, 12(193), pp. 1–27.

Cuellar, M. (2006). *¿A la vivienda quien la ronda?*, Instituto Colombiano de Ahorro y Vivienda, Bogota. Colombia: Universidad Externado de Colombia.

Dachary, A. C., & Arnaiz Burne, S. M. (2006). *El territorio y turismo. Nuevas dimensiones y acciones*. Guadalajara: Universidad de Guadalajara.

Dal Bó, E. (2006). Regulatory capture: A review. *Oxford Review of Economic Policy, 22(2)*, 203–225.

Dávila, A. (2016). *El mall: The spatial and class politics of shopping malls in Latin America*. Berkeley: University of California Press.

Dawson, J., Clarke, R., & Mukoyama, M. (2006). *Strategic issues in international retailing*. London: Routledge.

De la Torre, A., & Schmukler, S. L. (2007). *Emerging capital markets and globalization: The Latin American experience*. IDB Publications (Books), Inter-American Development Bank, number 349, December.

De Mattos, C. A. (2007). Globalización, negocios inmobiliarios y mercantilización del desarrollo urbano. *Nueva Sociedad, 212*(November–December).

De Simone, L. (2015). *MetaMall. Espacio urbano y consumo en la ciudad neoliberal chilena*. Santiago, Chile: Ril Editores-Pontificia/Universidad Católica de Chile.

DOF (Diario Oficial de la Federación) (2012, January 16). *Ley de Asociaciones Público Privadas*. Available at http://www.diputados.gob.mx/LeyesBiblio/pdf/LAPP_150618.pdf, accessed on April 18, 2019.

Duffie, D., & Thukral, M. (2012). *Redesigning credit derivatives to better cover sovereign default risk*. Stanford University Working Paper No. 3107.

Durand, C. (2009). *Wal-Mart en Mexico, una trayectoria exitosa y sus causas*. Paper presented at the International Seminar Metropolización, transformaciones mercantiles y gobernanza en los países emergentes las grandes ciudades en las mutaciones del comercio mundial, Colegio de México, July 1–3, 2009.

Durand, C. (2017). *Fictitious capital. How finance is appropriating our future*. London/New York: Verso.

ENIGH (2018). Encuesta Nacional de Ingresos y Gastos de los Hogares 2018. Available at https://www.inegi.org.mx/programas/enigh/nc/2018/.

Epstein, G. A. (2005). Introduction. In *Financialization and the World Economy* (pp. 3–16). Cheltenham: Edward Elgar.

Esquirol, J. L. (2008). Titling and untitled housing in Panama City. *Tennessee Journal of Law & Policy*, 4(2), 243–302.

Etulain, J. (2008). ¿Gestión promocional o privatización de la gestión urbanística? Proyecto urbano Puerto Madero, Buenos Aires – Argentina. *Bitácora Urbano Territorial*, 12(1), 171–184.

Euromonitor (2010). *Global hotels: Lagging but not lost*. Available at www.portal.euromonitor.com, accessed on March 28, 2018.

Evans, A. W. (1973). *The economics of residential location*. London and Basingstoke: Palgrave Macmillan.

Fabozzi, F. J., Paletta, T., Stanescu, S., & Tunaru, R. (2016). An improved method for pricing and hedging long dated American options. *European Journal of Operational Research*, 254(2), 656–666.

Fairfield, T. (2015). *Private wealth and public revenue in Latin America: Business power and tax politics*. New York: Cambridge University Press.

Fernandez, R., & Aalbers, M. B. (2019). Housing financialization in the Global South: In search of a comparative framework. *Housing Policy Debate*, 30(4), 680–701. DOI:10.1080/10511482.2019.1681491.

Fields, D. (2018). Constructing a new asset class: Property-led financial accumulation after the crisis. *Economic Geography*, 94(2), 118–140.

Financial Crisis Inquiry Commission (2010). *Lawrence Summers, interview by FCIC, May 28, 2010*. Veritext National Deposition & Litigation Services 866 299-5127. Washington, DC: Official Government Edition.

Financial Crisis Inquiry Commission (2011). *Crisis inquiry report, final report of the National Commission on the Causes of the Financial and Economic Crisis in the United States*. Washington, DC: Official Government Edition.

Fine, B. (2014). Financialization from a Marxist perspective. *International Journal of Political Economy*, 42(4), 47–66.

Fischer, T. (1996). Gestão contemporânea, cidades estratégicas: Aprendendo com fragmentos e reconfigurações do local. In T. Fischer (Ed.), *Gestão contemporânea: cidades estratégicas e organizações locais* (pp. 10–26). Rio de Janeiro: Editora da Fundação Getúlio Vargas.

Fix, M. (2001). Parceiros da exclusão: Duas histórias da construção de uma "nova cidade". In *São Paulo: Faria Lima e Agua Espraiada*. Boitempo Editorial. Available at https://books.google.be/books?id=tTtHAAAAYAAJ, accessed on May 3, 2019.

Flyvbjerg, B. (2014). What you should know about megaprojects and why: An overview. *Project Management Journal*, 45(2), 6–19.

Fourquet, François, & Murard, Lion (1978). *Los equipamientos del poder*. Barcelona: Editorial Gustavo Gili, S. A.

Foucault, M. (2012). *El poder, una bestia magnífica: Sobre el poder, la prisión y la vida* (Horacio Pons, Trans.). Buenos Aires: Siglo XXI.

Freedom House (2020). *Freedom in the world 2020*. Washington, DC: Freedom House. Available at https://freedomhouse.org/country/panama/freedom-world/2020, accessed on June 16, 2021.

Fujita, M., Krugman, P., & Venables, A. J. (1999). *The spatial economy. Cities, regions, and international trade*. Cambridge, MA: MIT Press.

Fundação João Pinheiro (2021). *Deficit habitacional no Brasil – 2016–2019*. Belo Horizonte: FJP.

Fundación Techo (2013). *Actualización del Catastro Nacional de Campamentos*. Santiago: Techo.

Furtado, B., Lima Neto, V., & Krause, C. (2013). *Estimativas do déficit habitacional brasileiro (2007–2011) por municipios (2010)*. Nota Técnica, 1. Brasília: IPEA.

G20, Leaders' Statement (2009, September 24–25). *The Pittsburgh Summit*, p. 8.

Gaceta Oficial del Distrito Federal (2000, September 12). *Decreto por el que se aprueba el Programa Parcial de Desarrollo Urbano de la Zona de Santa Fe*. Mexico City: Asamblea Legislativa Del Distrito Federal.

Gaceta Oficial del Distrito Federal (2012, May 4). *Decreto que contiene el Programa Parcial de Desarrollo Urbano de la "Zona Santa Fe", de los Programas Delegacionales de Desarrollo Urbano para las Delegaciones Álvaro Obregón y Cuajimalpa de Morelos*. Mexico City: Asamblea Legislativa Del Distrito Federal.

Ganster, M. (2001). Opportunities and challenges of investing in emerging markets: A case study of Panama. Thesis, Massachusetts Institute of Technology.

Garay, A. (2002). Acerca de la gestión de proyectos urbanos: Las enseñanzas de Puerto Madero. In J. Liernur (Ed.), *Puerto Madero Waterfront*. Collection CASE, N° 6. Cambridge, MA: Harvard University Graduate School of Design/Prestel.

Garay, A., Wainer, L., Henderson, H., & Rotbart, D. (2013). Puerto Madero: Análisis de un proyecto. *Land Lines*, July, Lincoln Institute of Land Policy.

Gasca-Zamora, J. (2017). Centros comerciales de la Ciudad de México: El ascenso de los negocios inmobiliarios orientados al consumo. *EURE*, 43(130), 73–96. Available at https://dx.doi.org/10.4067/s0250-71612017000300073, accessed on November 12, 2018.

Gómez-González, J. E., Ojeda-Joya, J. N., Rey-Guerra, C., & Sicard, N. (2013). *Testing for bubbles in housing markets: New results using a new method*. Federal Reserve Bank of Dallas, Globalization and Monetary Policy Institute Working Papers No. 164, pp. 1–11.

González, R., & Martínez Almazán, R. (Eds.) (2018). *Santa Fe. Una mirada hacia el futuro. Desarrollo urbano, gobernanza y administración pública*. Ciudad de México: Instituto Nacional de Administración Pública, A. C.

Gotham, K. F. (2006). The secondary circuit of capital reconsidered: Globalization and the U.S. real estate sector. *American Journal of Sociology*, 112(1), 231–275.

Gotham, K. F. (2009). Creating liquidity out of spatial fixity: The secondary circuit of capital and the subprime mortgage crisis. *International Journal of Urban and Regional Research*, 33(2), 355–371.

Grupo BMV (2012). *FIBRAS. Fideicomisos de infraestructura y bienes raíces.* México: BMV. Available at https://www.bmv.com.mx/docspub/MI_EMPRE SA_EN_BOLSA/CTEN_MINGE/Fibras.pdf, accessed on June 22, 2017.

Harris, C. D., & Ullman, E. L. (1945). The nature of cities. *Annals of the American Academy of Political and Social Sciences,* 24(2), 7–17.

Harvey, D. (1973). *Social justice and the city.* Baltimore, MD: John Hopkins University Press.

Harvey, D. (1982). *The limits to capital.* Oxford: Blackwell.

Harvey, D. (1989a). *The condition of postmodernity: An inquiry into the origins of cultural change.* Oxford: Basil Blackwell.

Harvey, D. (1989b). From managerialism to entrepreneurialism: The transformation in urban governance in late capitalism. *Geografiska Annaler: series B, human geography,* 71(1), 3–17.

Harvey, D. (1999). *The limits to capital.* London/New York: Verso.

Harvey, D. (2001). *Spaces of capital: Towards a critical geography.* Edinburgh: Edinburgh University Press.

Harvey, D. (2003). *The new imperialism.* Oxford: Oxford University Press.

Harvey, D. (2004). The 'new' imperialism: Accumulation by dispossession. *Socialist Register,* 40, 63–87.

Harvey, D. (2007). *Breve historia del neoliberalismo.* Madrid: Akal.

Harvey, D. (2010). *The enigma of capital and the crises of capitalism.* London: Profile Books.

Harvey, D. (2014). *Seventeen contradictions and the end of capitalism.* London: Profile Books.

Harvey, D. (2019). *Marx, el capital y la locura de la razón económica.* CDMX: Akal.

Harvey, D., & Rivera, H. A. (2010). Explaining the crisis. *International Socialist Review,* 73. Available at https://isreview.org/issue/73/explaining-crisis/, accessed on February 15, 2022.

Hawkins, D. (2009). *El turismo como una estrategia de asistencia para el desarrollo.* Ponencia I Congreso de Turismo, Cooperación y Desarrollo, 15–16 octubre 2009, Vilaseca (Tarragona, España).

Hidalgo, R., Azar, P., Borsdorf, A., & Paulsen, A. (2016). Hospedándose en la ciudad global: Patrones de localización de los hoteles de lujo en Santiago de Chile. *Cuadernos de Geografía: Revista Colombiana de Geografía,* 25(2), 221–236.

Honey, M. (2008). *Ecotourism and sustainable development: Who owns Paradise?* (2nd ed.). Washington, DC: Island Press.

Honohan, P. (2008). Cross-country variation in household access to financial services. *Journal of Banking & Finance,* 32(11), 2493–2500.

Hoover, E. M., & Vernon, R. (1959). *Anatomy of a metropolis: The changing distribution of people and jobs within the New York metropolitan region.* Cambridge, MA: Harvard University Press.

Hosteltour (2009). *El 60% de planta hotelera de Quintana Roo es de origen español, porcentaje que va en aumento.* Available at http://www.hosteltur.com/noticias/ 58509_60-planta-hotelera-quintana-roo-es-origen-espanol-porcentaje-va-aumento.html, accessed on February 5, 2011.

Hoyt, H. (1939). *The structure and growth of residential neighborhoods in American cities.* Washington, DC: Government Printing Office.

Hoyt, H. (1964). Recent distortions of the classical models of urban structure. *Land Economics*, 40(2), 282–295.

IADB (Inter-American Development Bank) (2020). *Financing sustainable infrastructure in Latin America and the Caribbean. Market development and recommendations*. Washington, DC: Inter-American Development Bank.

IBGE (Instituto Brasileiro de Geografia e Estatistica) (2006). *Pesquisa anual da industria da construção*. Rio de Janeiro: IBGE.

ICSC (International Council of Shopping Centers) (1999). *ICSC shopping center definitions: Basic configurations and types*. Available at https://eduardoquiza.files. wordpress.com/2009/09/scdefinitions99.pdf, accessed on July 23, 2011.

ICSC (International Council of Shopping Centers) (2015). *Reporte 2015 de la industria de centros comerciales en América Latina*. México: International Council of Shopping Centers. Available at https://www.icsc.com/uploads/ event_presentations/Reporte_de_la_industria-_Industry_Report_2015.pdf, accessed on May 12, 2011.

IDOM-SUMA-CONTRANS (2018) *Memoria plan parcial de ordenamiento territorial de San Francisco, distrito y provincia de Panamá*. Panama City: Alcaldía del Distrito de Panamá.

IMF (International Monetary Fund) (2008). *Global financial stability report: Financial stress and deleveraging, macro financial implications and policy*. Washington, DC: International Monetary Fund.

IMF (International Monetary Fund) (2014). *Executive board report, 2014*. Available at https://www.imf.org/external/np/sec/pr/2014/pr14235.htm.

Imilan Ojeda, W. (2016). *Políticas y luchas por la vivienda en Chile: El camino neoliberal*. Working paper series Contested Cities. Serie (V) Políticas y luchas por la vivienda. WPCC-16004, 21 pp.

Immergluck D., & Law, J. (2014). Speculating in crisis. The intrametropolitan geography of investing in foreclosed homes in Atlanta. *Urban Geography*, 35(1), 1–24.

INEC (2015). *Panamá en Cifras, Instituto Nacional de Estadística y Censo. 2010–2014*. Ciudad de Panamá: Contraloría General de la República de Panamá.

INEGI (2014). *Resultados de la Encuesta Nacional de Ocupación y Empleo (ENOE). II trimestre de 2014*. Available at http://www3.inegi.org.mx/sistemas/tabulados basicos/tabtema. aspx?s=est&c=33619, accessed on February 15, 2022.

INEGI (2020). *2020 Census of Population and Housing*. Aguascalientes: INEGI. Available at http://en.www.inegi.org.mx/programas/ccpv/2020/#:~:text= With%20the%202020%20Census%2C%20the,been%20carried%20 out%20since%201895.&text=In%201892%20a%20pilot%20populat ion,City%2C%20called%20the%20Pe%C3%B1afiel%20Census, accessed on September 12, 2021.

INFONAVIT (2014). *Atlas del abandono de vivienda*. CDMX: Instituto del Fondo Nacional de la Vivienda para los Trabajadores.

INFONAVIT (2019). *Reporte Anual Vivienda INFONAVIT*. CDMX: Infonavit. Available at https://portalmx.infonavit.org.mx/wps/wcm/connect/6a22332f- f9fe-4f17-8d93-9efc959086b2/ReporteAnualVivienda2019.pdf?MOD=AJPE RES&CVID=mW5tCKM, accessed on February 15, 2022.

INFONAVIT (2020). *Reporte económico trimestral enero-marzo 2020, No.4, Gerencia de estudios económicos de Infonavit*. CDMX: Infonavit.

Innes, A. (2014). The political economy of state capture in Central Europe. *Journal of Common Market Studies*, 52(1), 88–104.

Irazábal, C. (2018). Coastal urban planning in 'The Green Republic': Tourism development and the nature-infrastructure paradox in Costa Rica. *International Journal of Urban and Regional Research*, 42(5), 882–913. Available at https://doi.org/10.1111/1468-2427.12654, accessed on February 15, 2022.

Isa Contreras, P. (2011). Expansión y agotamiento del modelo turístico dominicano. El turismo en los informes de desarrollo humano en la República Dominicana. In E. Cañada & M. Blázquez (Eds.), *Turismo placebo: Nueva colonización turística: Del Mediterráneo a Mesoaméricas y El Caribe. Lógicas espaciales del capital turístico* (pp. 11–28). Managua: Edisa.

ITC/WTO (2015). *Tourism and trade: A global agenda for sustainable development*. Geneva: International Trade Centre and World Tourism Organization.

Ito, K. (2008). Global imbalances: Origins, consequences and possible resolutions. In J.-P. Touffout (Ed.), *Central banks as economic institutions* (pp. 142–161). Cheltenham: Edward Elgar.

Jacobs, J. (1970). *The economy of cities*. London: Jonathan Cape.

Jaramillo González, E. S., & Cuervo Ballesteros, N. (2014). *Precios inmobiliarios de vivienda en Bogotá 1970–2013*. Documentos CEDE 18. Universidad de los Andes.

Jaramillo González, S. (1994). *Hacia una teoría de la renta del suelo urbano*. Bogotá: Universidad de los Andes.

Jaramillo González, S. (2009). *Hacia una teoría de la renta del suelo urbano*. Bogotá: Ediciones Uniandes.

Jessop, B. (2004). La economía política de la escala y la construcción de regiones transfronterizas. *EURE*, 29(89), 25–41.

Kaufmann, D., & Hellman, J. (2001). Confronting the challenge of state capture in transition economies. *Finance & Development*, 38(3).

Knafou, R. (2006). El turismo, factor de cambio territorial: Evolución de los lugares, actores y prácticas a lo largo del tiempo (del s. XVIII al s. XXI). In A. Lacosta (Ed.), *Turismo y cambio territorial: ¿eclosión, aceleración, desbordamiento?* IX Coloquio de Geografía del Turismo, Ocio y Recreación. Zaragoza: Prensas Universitarias de Zaragoza.

Kothari, V. (2006). *Securitization. The financial instrument of the future*. Singapore: John Wiley & Sons.

Krippner, G. (2005). The financialization of the American economy. *Socio-Economic Review*, 3, 173–208.

Lapavitsas, C. (2013). The financialization of capitalism: 'Profiting without producing'. *City*, 17(6), 792–805.

Lascoumes, P., & Le Galès, P. (2007). Introduction: Understanding public policy through its instruments – from the nature of instruments to the sociology of public policy instrumentation. *Governance*, 20(1), 1–21.

Lawrence, J. T. (2017). Following the money: Lessons from the Panama Papers, Part 1: Tip of the iceberg. *Dickinson Law Review*, 121(3), 807–874.

Lefebvre, H. (1970). *La révolution urbaine*. Paris: Éditions Gallimard.

Lefebvre, H. (1973). *De lo rural a lo urbano*. Barcelona: Peninsula.

Lefebvre, H. (1974). *La production de l'espace*. Paris: Anthropos.

Lefebvre, H. (1976). *Espacio y política*. Barcelona: Peninsula.

León-Darder, F., Villar-Garcia, C., & Pla-Barber, J. (2011). Entry mode choice in the internationalization of the hotel industry: A holistic approach. *Service Industries Journal*, 31, 107–122.

Lima, T. A., Sene, G. M., & Torres de Souza, M. A. (2016). Em busca do Cais do Valongo, Rio de Janeiro, século XIX. *Anais do Museu Paulista*, 24(1), 299–390.

Lipietz, A. (1974). *Le tribut foncier urbain*. Paris: Maspero.

Lipietz, A. (1985). A Marxist approach to urban ground rent: The case of France. In Michael Ball et al. (Eds.), *Land rent, housing and urban planning* (pp. 129–155). London: Croom Helm.

LISR (2016). *Ley del impuesto sobre la renta*. Mexico City: Cámara de Diputados.

Littlejohn, D. (2003). Hotels. In B. Brotherton (Ed.), *The international hospitality industry: Structure, characteristics and issues* (pp. 5–29). Oxford: Butterworth-Heinemann.

Lizán, J. (2014). Los centros comerciales y su futuro 'promisorio'. *Obras web/ Inmobiliario*. Available at http://www.obrasweb.mx/inmobiliario/2014/10/09/la-evolucion-de-los-centroscomerciales-y-su-futuro-promisorio, accessed on June 12, 2019.

López, L. (2006). Centros comerciales, recintos fortificados. *Veredas*, 7(12), 147–163. Available at http://bidi.xoc.uam.mx/tabla_contenido_fasciculo.php?id_fasciculo=268, accessed on April 7, 2011.

Lundberg, D., Krishnamoorthy, M., & Starvey, M. (1995). *Tourism economics*. New York: John Wiley & Sons.

Marazzi, C. (2009). La violencia del capitalismo financiero. In A. Fumagalli, S. Lucarelli, A. Negri, & C. Vercellone, *La gran crisis de la economía global. Mercados financieros, luchas sociales y nuevos escenarios políticos* (pp. 21–61). Madrid: Traficantes de Sueños.

Markowitz, H. (1952). Portfolio selection. *Journal of Finance*, 7(1), 77–91.

Markowitz, H. M. (1959). *Portfolio selection: Efficient diversification of investments*. New York: John Wiley & Sons.

Marichal, C. (2010). *Nueva historia de las grandes crisis financieras. Una perspectiva global 1873–2008*. Buenos Aires: Debate.

Marosi, R. (2017, November 26). A subprime horror. *Los Angeles Times*.

Marx, K. (1971). *Fundamentos para crítica de la economía política*. La Habana: Editorial de Ciencias Sociales.

Marx, K. (1973). *El capital*. México: FCE.

Marx, K. (1976). *Capital – A critique of political economy*, Vol. 1. Harmondsworth: Penguin Books, in association with New Left Review.

Marx, K. (1978). Manifiesto del partido comunista. In *Obras de Marx y Engels*, Vol. 9. Barcelona: Crítica.

Marx, K. (1998). *El capital*. Ciudad de México: Siglo XXI.

Marx, K. (2007). *Elementos fundamentales para la crítica de la economía política (Grundrisse 1857–1858)*. 3 Vols. México: Siglo XXI.

Marx, K., & Engels, F. (2004). *Karl Marx and Frederick Engels: Collected works*. Volumes 1–50. New York: International Pub.

Marx, K., & Engels, F. (2010). *Marx & Engels. Collected Works*, Vol. 37. *Karl Marx Capital*, Vol. 3. London: Lawrence & Wishart Electric Book.

McKenzie, R. (1924). The ecological approach to the study of the human community. *American Journal of Sociology*, 30(3), 287–301.

McKenzie, R. D. (Ed.) (1933). *Metropolitan community*. New York: McGraw-Hill.

MEF/BM (2008). *Encuesta de niveles de vida 2008*. Panama: Ministerio de Economía y Finanzas (MEF). Available at https://microdata.worldbank.org/index.php/catalog/70, accessed on March 23, 2017.

Ministerio de Desarrollo Social (2013). *Informe de Política Social 2013*. Santiago de Chile.

Minsky, H. P. (1982). *Can "it" happen again?: Essays on instability and finance*. New York: M.E. Sharpe.

MINVU (1979). *Politica nacional de desarrollo urbano*. Santiago de Chile: MINVU.

MINVU (2009). *Informe final de evaluación programa subsidio leasing habitacional*. Santiago de Chile: MINVU.

MINVU (2014). *Catastro nacional de condominios sociales*. Santiago de Chile: Secretaría Ejecutiva de Recuperación de Barrios.

MINVU (2016). *Modifica Decreto Supremo N°14 (V. y U.), de 2007, que aprueba reglamento del Programa de Recuperación de Barrios, en el sentido que indica*. Available at https://www.leychile.cl/Navegar?idNorma=1103190, accessed on February 15, 2022.

MINVU (2018). *Informe de Gestión Ministerial Mejoramiento y Regeneración de Condominios de Viviendas 2014–2018*. Santiago de Chile: MINVU. Available at https://www.senado.cl/site/presupuesto/2018/cumplimiento/Glosas%202018/cuarta%20subcomision/18%20Vivienda/183%20Vivienda.pdf, accessed on February 15, 2022.

Moody's (2016). *Situación del mercado y riesgo operativo en finanzas estructuradas*. https://www.moodys.com/sites/products/ProductAttachments/Evoluci%C3%B3n%20y%20Desempe%C3%B1o%20del%20Mercado%20Mexicano.pdf, accessed on February 15, 2022.

Mohan, R. (1979). *Urban economic and planning models. Assessing the potential for cities in developing countries*. World Bank Staff Occasional Papers No. 25. Baltimore & London: World Bank/The Johns Hopkins University Press.

Monkkonen, P. (2011). The housing transition in Mexico: Expanding access to housing finance. *Urban Affairs Review*, 47(5), 672–695.

Moro, Vinícius (2011). *Projeto Porto Maravilha*. Rio de Janeiro: Rio Prefeitura.

Mosciaro, M., & Santos Pereira, A. L. (2017). *Reinforcing uneven development: The financialization of Brazilian urban redevelopment projects*. Working paper.

Moses, L. N., & Williamson, H. F. (1967). The location of economic activity in cities. *American Economic Review*, 57(2), 211–222.

Mulligan, G. F., Partridge, M. D., & Carruthers, J. I. (2012). Central place theory and its reemergence in regional science. *Annals of Regional Science*, 48, 405–431.

NAREIT (National Association of Real Estate Investment Trusts) (1993, February). *REIT Watch*. Washington, DC.

Navarro, E., Thiel, D., & Romero, Y. (2015). Periferias del placer: Cuando turismo se convierte en desarrollismo inmobiliario-turístico. *Boletín de la Asociación de Geógrafos Españoles*, 67(1), 275–302.

O Globo (2020). *Zona Portuária, uma região de contrates*. Agência O Globo. Available at https://oglobo.globo.com/rio/zona-portuaria-uma-regiao-de-contrastes-24210821, accessed on January 3, 2021.

Obermayer, B., & Obermaier, F. (2016). *The Panama Papers: Breaking the story of how the rich and powerful hide their money*. London: Simon and Schuster.

OECD (2015). *Estudios de la OCDE sobre los sistemas de pensiones: México.* Paris: Organización para la Cooperación y el Desarrollo Económicos.

OECD et al. (2019). *Latin American Economic Outlook 2019: Development in Transition.* Paris: OECD Publishing. Available at https://doi.org/10.1787/g2g9f f18-en. accessed on April 6, 2020.

Office of Investor, Education and Advocacy (2011). Investor Bulletin: Real Estate Investment Trusts (REITs). *Investor Assistance,* 800, 1–5.

Oliveira, F. B. de, Sant'Anna, A. de S., Diniz, D. M., & Carvalho Neto, A. M. de (2015). Leaderships in urban contexts of diversity and innovation: The Porto Maravilha case. *Brazilian Administration Review,* Rio de Janeiro, 12(3), 268–287.

Ortalo-Magne, F., & Rady, S. (1999). Boom in, bust out: Young households and the housing price cycle. *European Economic Review,* 43(4–6), 755–766. Available at https://doi.org/10.2139/ssrn.137961, accessed on February 15, 2022.

O'Sullivan, A. (2012). *Urban economics.* New York: McGraw-Hill/Irwin.

Park, R. E., Burgess, E. W., & McKenzie, R. D. (2019) [1925]. *The city. Suggestions for investigation of human behavior in the urban environment.* Chicago: University of Chicago Press.

Parsons, T. (1951). *The social system.* Glencoe, IL: The Free Press.

Pereira, A. (2017). Financialization of housing in Brazil: New frontiers. *International Journal of Urban and Regional Research,* 41(4), 604–622. Available at https://doi.org/10.1111/1468-2427.12518, accessed on February 15, 2022.

Pérez Negrete, M. (2017). *Megaproyectos, capital y resistencias: Una mirada desde la antropología urbana.* México City: CIESAS.

Piketty, T. (2014). *Capital in the twenty-first century.* Cambridge, MA: Harvard University Press.

Polanyi, K. (2001) [1944]. *The great transformation: The political and economic origins of our time.* Boston, MA: Beacon Press.

Porter, M. E. (1995). The competitive advantage of the inner city. *Harvard Business Review,* May–June, 55–71.

Prabha, A., Savard, K., & Wickramarachi, H. (2014). *Deriving the economic impact of derivatives. Growth through risk management.* Santa Monica, CA: Milken Institute.

Prefeitura da Cidade do Rio de Janeiro (2017). *Prefeito apresenta Plano de Habitação de Interesse Social do Porto.* Rio de Janeiro, October 1st. Available at: http://www.rio.rj.gov.br/web/guest/exibeconteudo?id=5621886, accessed on June 24, 2021.

Puebla, C. (2002). *Del intervencionismo estatal a las estrategias facilitadoras: Cambios en la política de vivienda en Mexico (1972–1994).* Ciudad de México: El Colegio de Mexico.

Rajan, R. (2005). Has financial development made the world riskier? National Bureau of Economic Research, Working Paper 11728, pp. 313–369. Available at https://www.nber.org/system/files/working_papers/w11728/w11728.pdf, accessed on February 15, 2022.

Razmilic, S. (2010). *Property values, housing subsidies and incentives: Evidence from Chile's current housing policies.* Master's dissertation, Massachusetts Institute of Technology.

Reardon, T., Henson, S., & Berdegué, J. (2007). Proactive fast-tracking diffusion of supermarkets in developing countries: Implications for market institution and trade. *Journal of Economic Geography*, 7, 399–431.

Rémy, J. (1987). La crise de la professionnalisation en agriculture: Les enjeux de la lutte pour le contrôle du titre d'agriculteur. *Sociologie du travail*, 87(4), 415–441.

Retail Traffic (2010). The Top Ten. *Retail Traffic*, June.

Rodríguez, A. (1995). Planificando la revitalización de una vieja ciudad industrial: Innovaciones de la política urbana en Bilbao Metropolitana. *Revista Interamericana de Planificación*, 110, Cuenca, Ecuador.

Rodríguez, A., & Icaza, A. (1993). Procesos de expulsión de habitantes de bajos ingresos del centro de Santiago, 1981 1990. *Proposiciones*, 22, 138–172. Available at http://www.sitiosur.cl/r.php?id=225, accessed on February 15, 2022.

Rodríguez, A., & Sugranyes, A. (Eds.) (2005). *Los con techo. Un desafío para lapolítica de vivienda social.* Santiago: Ediciones SUR.

Rodríguez, A., & Sugranyes, A. (2012). El traje nuevo del emperador. Las políticas de financiamiento de vivienda social en Santiago de Chile. In Jaime Erazo (Ed.) *Políticas de empleo y vivienda en Sudamérica* (pp. 47–73). Quito: FLACSO/ CLACSO.

Rolnik, R. (2015). *A guerra dos lugares: A colonização da terra e da moradia na era das finanças.* São Paulo: Boitempo.

Royer, L. (2009). *Financeirização da política habitacional: Limites e perspectivas* [Financialization of the housing policy: Limits and perspectives]. PhD dissertation, University of São Paulo.

Royer, L. (2014). *Financeirização da politica habitacional. Limites e perspectivas.* São Paulo: Annablume.

Salcedo-Hansen, R. (2003). Lo local, lo global y el mall: La lógica de la exclusión y la interdependencia. *Revista de Geografía Norte Grande*, 30, 103–115. Available at http://www.redalyc.org/articulo.oa?id=30003009, accessed on January 29, 2011.

Sandroni, P. (2010). A new financial instrument of value capture in São Paulo: Certificates of additional construction potential. In G. K. Ingram & Y. H. Hong (Eds.), *Municipal revenues and land policies* (pp. 218–236). Cambridge, MA: Lincoln Institute of Land Policy.

Sanfelici, D. (2013). La financiarización y la producción del espacio urbano en Brasil: Una contribución al debate. *Revista EURE – Revista de Estudios Urbano Regionales*, 39(118). Available at http://www.eure.cl/index.php/eure/article/ view/408/610, accessed on February 15, 2022.

Santos, M. (1977). Spatial dialectics: The two circuits of urban economy in underdeveloped countries. *Antipode*, 9(3), 49–60.

Sassen, S. (1991). *The global city: New York, London, & Tokyo* (2nd ed.). Princeton, NJ: Princeton University Press.

Sassen, S. (2012). Expanding the terrain for global capital. When local housing becomes an electronic instrument. In Manuel B. Aalbers (Ed.), *Subprime cities: The political economy of mortgage markets* (pp. 74–96). London: Blackwell.

Sassen, S. (2014). *Expulsions: Brutality and complexity in the global economy.* Cambridge, MA: Harvard University Press.

Saunders, P. (1986). *Social theory and the urban question*. New York: Holmes & Meier.

Schopflocher, T., & Manzi, J. M. (2020). Global structured finance outlook 2020: Another $1 trillion-plus year on tap. *S & P Global Ratings*. Available at https://www.spglobal.com/ratings/en/research/articles/200106-global-str uctured-finance-outlook-2020-another-1-trillion-plus-year-on-tap-11301282, accessed on May 3, 2019.

Schumpeter, J. A. (1942). *Capitalism, socialism, and democracy*. New York: Harper & Row.

Schumpeter, J. A. (1983) [1942]. *Capitalismo, socialismo y democracia*. Barcelona: Orbi.

Schwartz, H. M. (2009). *Subprime nation: American power, global finance, and the housing bubble*. Ithaca, NY: Cornell University Press.

SEDUVI (2003). *Zonificación y normas de ordenación*. CDMX: Gobierno del Distrito Federal.

Serebrisky, T., Suárez-Alemán, A., Margot, D., & Ramirez, M. (2015). *Financing infrastructure in Latin America and the Caribbean: How, how much and by whom?* Washington, DC: Inter-American Development Bank. Available at https://publications.iadb.org/publications/english/document/, accessed on June 2, 2011.

Siqueira, M. T. (2014). Entre o fundamental e o contingente: Dimensões da gentrificação contemporânea nas operações urbanas em São Paulo. *Cadernos Metrópole*, 16(32), 391–415.

Simmel, G. (1903). The metropolis and mental life. In Gary Bridge & Sophie Watson (Eds.), *The Blackwell city reader* (pp. 103–110). Oxford & Malden, MA: Wiley-Blackwell, 2002.

Smith, N. (1996). *The new urban frontier: Gentrification and the revanchist city*. New York: Routledge.

Smolka, M. O. (2013). *Implementación de la recuperación de plusvalías en América Latina: Políticas e instrumentos para el desarrollo urbano*. Cambridge: Lincoln Institute of Land Policy.

Socoloff, I. (2015). Financiamiento global y centros comerciales en Buenos Aires: Un estudio del caso IRSA. *Revista INVI*, 30(84), 151–177. Available at https://dx.doi.org/10.4067/S0718-83582015000200006, accessed on June 22, 2019.

Soja, E. (2000). *Postmetropolis critical studies of cities and regions*. Oxford: Blackwell.

Solo, T. M., & Manroth, A. (2006). *Access to financial services in Colombia*. Policy Research Working Paper 3834, World Bank, Washington, DC.

Sommi, L. (2005). La crisis de 1929 en América Latina. *Iztapalapa*, 91–100.

Soto, R. (2013). América Latina. Entre la financiarización y el financiamiento productivo. *Revista Problemas del Desarrollo*, 44(173). Available at https://doi.org/10.1016/S0301-7036(13)71875-3, accessed on February 15, 2022.

Souty, J. (2013). Dinâmicas de patrimonialização em contexto de revitalização e de globalização urbana. Notas sobre a Região Portuária do Rio De Janeiro. *Revista Memória em Rede*, 5(9). Available at http://www2.ufpel.edu.br/ich/memoriaemrede/beta-02-01/index.php/memoriaemrede/article/view/211/0, accessed on February 15, 2022.

Spolon, A. P. (2008). Hospitalidade contemporanea nas grandes cidades da América Latina: Sao Paulo, Santiago e Buenos Aires em foco. In Cesar Javier

Pereira & Rodrigo Hidalgo (Eds.), *Producción inmobiliaria y reestructuración metropolitana en América Latina* (pp. 323–340). Santiago: Universidade do Sao Paulo and Universidad Católica de Chile.

Standard & Poor's (2015). *Global infrastructure investment: Timing is everything (and now is the time)*. Standard & Poor's Ratings Services, McGraw Hill Financial. Available at https://www.haberhabere.com/english-news/sp-global-infrastructure-investment-timing-is-everything-and-now-is-the-time-h3687.html, accessed on August 3, 2021.

Standard & Poor's Financial Services LLC (2020). *Latin American structured finance 2020 outlook*. Available at https://www.spglobal.com/ratings/en/index, accessed on February 15, 2022.

Stiglitz, J. E. (2001). Foreword. In *The great transformation: The political and economic origins of our time*. Boston, MA: Beacon Press.

Stiglitz, J. E. (2012). *The price of inequality: How today's divided society endangers our future*. New York: W.W. Norton.

Story, L. (2005, June 15). Buyout firm to acquire Wyndham. *The New York Times*. Available at https://www.nytimes.com/2005/06/15/business/buyout-firm-to-acquire-wyndham.html, accessed on October 18, 2021.

Swyngedouw, E., Moulaert, F., & Rodríguez, A. (2002). Neoliberal urbanization in Europe: Large-scale urban development projects and the new urban policy. *Antipode*, 34(3), 542–577.

Tavakoli, J. M. (2008). *Structured finance and collateralized debt obligations: New developments in cash and synthetic securitization* (Vol. 509). Hoboken, NJ: John Wiley & Sons.

Taylor, P. J. (2004). *World city network: A global urban analysis*. London: Routledge.

The Economist (2009, June 18). No empty threat: Credit-default swaps are pitting firms against their own creditors.

Thomas, L. C. (2000). A survey of credit and behavioural scoring: Forecasting financial risk of lending to consumers. *International Journal of Forecasting*, 16, 149–172.

Thünen, J. H. von (1966). *Von Thünen's isolated state; an English edition of Der isolierte Staat* (P. Hall, Ed.). Oxford/New York: Pergamon Press.

Titularizadora Colombiana (2005). *Financiación de vivienda y titularización hipotecaria en Latinoamérica*. Bogotá.

Topalov, C. (1974). *Les promoteurs immobiliers*. Paris: Mouton.

Topalov, C. (1979). *La urbanización capitalista*. CDMX: Edicol.

Topalov, C. (1990). Hacer la historia de la investigación urbana. La experiencia francesa desde 1965. *Sociológica*, 5(12), 175–207.

Transparency International (2016). *Unmask the corrupt*. Berlin: Transparency International. Available at https://unmaskthecorrupt.org, accessed on July 20, 2021.

Ullman, E. L. (1941). A theory of location for cities. *American Journal of Sociology*, 46, 853–864.

Ullman, E. L. (1962). Presidential address: The nature of cities reconsidered. *Papers and Proceedings of the Regional Science Association*, 9(1), 7–23.

UN-Habitat (2009). *Housing finance mechanisms in Chile*. The Human Settlements Finance Systems Series. Nairobi: United Nations Human Settlements Programme.

UN-Habitat (2013). *Scaling-up affordable housing supply in Brazil: The "My House My Life" programme.* Nairobi: United Nations Human Settlements Programme.

Valenzuela Aguilera, A. (2013). Dispositivos de la globalización: La construcción de grandes proyectos urbanos en Ciudad de México. *Revista EURE – Revista de Estudios Urbano Regionales*, 39(116), 101–118. Available at http://www.eure. cl/index.php/eure/article/view/258, accessed on July 23, 2021.

Valenzuela Aguilera, A. (2017). Failed markets: The crisis in the private production of social housing in Mexico. In *Latin American Perspectives*, Special issue on Housing, Infrastructure and Inequality in Latin American Cities, 44(2), 38–51.

Valenzuela Aguilera, A., & Tsenkova, S. (2019). Build it and they will come: Whatever happened to social housing in Mexico. *Urban Research & Practice*, 12(4), 493–504.

Walks, R. A. (2010). Bailing out the wealthy: Responses to the financial crisis, Ponzi neoliberalism, and the city. *Human Geography*, 3(3), 54–84.

Warf, B. (2002). Tailored for Panama: Offshore banking at the crossroads of the Americas. *Geografiska Annaler: series B, human geography*, 84(1), 33–47.

Weber, A. (1958 [1929]). *Theory of the location of industries.* Chicago: University of Chicago Press.

Weber, R. (2010). Selling city futures: The financialization of urban redevelopment policy. *Economic Geography*, 86(3), 251–274.

Wirth, L. (1988 [1938]). El urbanismo como modo de vida. In Mario Bassols & Roberto Donoso (Eds.), *Antología de sociología urbana* (pp. 162–182). México: UNAM.

World Bank (1993). *Housing: Enabling markets to work.* A World Bank policy paper. Washington, DC: World Bank.

World Bank (2021). *World Bank development indicators.* Washington, DC: World Bank. Available at https://tradingeconomics.com/indicators, accessed on February 21, 2021.

Wrigley, N., & Lowe, M. S. (2007). Transnational retail and the global economy. *Journal of Economic Geography*, 7, 337–341.

Young, T., McCord, L., & Crawford, P. J. (2010). Credit default swaps: The good, the bad and the ugly. *Journal of Business & Economics Research*, 8(4), 29–36.

Zhang, Xing Quan, & da Rocha Lima, Jr, João (2010). *Housing finance mechanisms in Brazil.* Nairobi: United Nations, UN-Habitat.

Index

absolute rent 40
accessibility 12, 18, 20, 28, 40–1, 43, 58–9
accumulation by dispossession 38, 80, 181
affordable housing 29, 104, 188–9
agglomerations 5, 10, 12, 15, 32–3, 58, 163
Airbnb 184
alienation 49
Alonso, William 8, 13
amenities 12, 25, 28, 31, 39, 40, 42, 58, 86, 101, 103, 118, 137, 155–6, 161–2, 178
American Depositary Receipts 61
American International Group (AIG) 76
appreciation 10, 55, 64, 124, 132–3, 143, 155, 161, 174, 192, 198; of land prices 43, 45, 197; of land value 12, 17, 34, 40–1, 46, 54, 57, 148, 151, 191; of property 28
arbitrage 73–5, 79, 160
Asset-Backed Securities 5, 70, 72, 77, 79, 144; complex financial structures 79; financial instruments 2, 33, 37, 63, 163, 170, 176, 180; securitization 3, 5, 65–79, 93–5, 105, 111, 114, 118–19, 126–7, 170, 178, 188–90

bailout 78, 80, 82, 85, 121, 124
Banco Estado 98–9
BANOBRAS 105, 170
BANXICO 107
Basel III framework 111
bespoke tranche opportunities 5
betterment contributions 57; taxation of development value 57
betterment levies, taxes levied 57
Bolsa de Comercio de Santiago 98
bonds 35, 61, 67, 74, 81, 83–4, 94–5, 126–7, 154–5, 165–70, 190, 193–4
boom and bust 38, 103
BORHIS 105

Borja, Jordi 135
Bourdieu, Pierre 51–2, 54
Brazil 1, 3, 52, 69, 87, 93–4, 96, 113–19, 127, 164, 167, 177, 188, 190
bubbles 38, 91, 125; housing bubbles 31, 113, 124–5; market bubbles 86; real estate bubbles 27–8, 40, 161
Buenos Aires 22, 132–3, 134, 162, 179, 192
Buffet, Warren 79, 160
building density 47
building stock 10, 197
built environment 9, 40, 65, 102–3, 163, 165, 188, 192
Burgess, Ernest, W. 16–20
business centers 26, 33, 65
business cycles 54
business districts 3, 63, 130, 132, 135, 162; central business district 21–2, 27, 29, 48, 58, 86; commercial activities 58; corporation headquarters 58; financial services 126, 170; functional specialization 10, 15, 27; housing developers 48, 102, 104–5, 111; market values see also land values

Cadernetas de Poupança 118
Caixa Econômica Federal (CAIXA) 116–19
capital accumulation 11, 36–7, 66, 103, 116, 160–2, 175, 181, 192, 196–7
capital circulation 37, 43, 183, 197
capital development certificates (CKDs)169–72, 178, 195; AFOREs 169–70, 174; estate investment 1–2, 5, 60, 66, 70, 73, 83–4, 133, 136, 141, 146, 154, 162, 166, 173, 182, 184, 195, 198; see also real estate; infrastructure bonds 169–70; see also infrastructure

capital fixation, capital gains 2, 10, 27, 39–40, 43, 46–7, 54–6, 59, 64, 131–3, 136, 141, 143, 148, 161–2, 169, 174, 185n8, 186, 191

capital investment 2, 5–6, 8, 10, 27–8, 33–4, 53, 70, 78, 88, 92, 186, 198

Capital Markets Law, Stock Market Law 172

capital valorization 35

capital value 37, 59

capitalist mode of production 8–9, 57

capitalization of rent 39, 41, 55

capture of the State 56

Cartas de Crédito, credit card 74, 179, 180, 195

cash CDOs 73, 75

cash-flow waterfalls 79

CDOs squared 76

CDSs-related assets, CDSs 69, 72–6, 78–82, 185n3

central places theory, central places systems 14

CEPACs 154–6, 193

CERPIS 169, 171

Costanera 133, 135–6, 157n1

certificates 35, 53, 88, 103, 111, 114, 119, 144, 148, 159, 165–6, 169, 170–5, 190, 195

certificates of real estate 114, 118, 173; receivables 61, 74, 118; revenues 33, 35, 36, 114, 124, 136–9, 143–4, 150, 154–7, 159, 163, 165, 172–4, 179, 181–3, 196

Chicago School 15–18

Christaller, Walter 14, 15, 22, 48–50

citizen participation certificates 165

city center 13, 25, 32, 184

city structure 5, 7, 20, 33, 186

collateral 36–8, 72, 75–82, 87, 91, 92, 95, 120, 126–7, 148

collateralized debt obligations 5, 38, 72, 76

Collateralized debt positions 5, 38, 72–3

collective property 45

Colombia 1, 3, 52, 94–6, 119–27, 146, 167, 190

commercial areas 3

commercial centers 186

commercial mortgage-backed securities 88n6; commercial mortgage loans 74

commercial real estate 65, 83, 86–7, 88n1, 133, 179; corporate business 65, 130; industries 7, 11, 15, 24, 32, 138, 150, 152, 173, 179, 193; office centers

21; office space 86, 151, 157; office buildings 87, 194; shopping centers 3, 22, 27–8, 30–1, 40, 63, 65, 86, 135, 161–2, 173, 176–9, 195

commercial strips 3

commodification 11, 97, 156

commodities 7, 14, 35–6, 39, 40, 61n1, 62n3, 67, 103

Commodities Futures Trading Commission 67

communication networks 11

community center 86, 177, 180

compensation 57, 78, 174, 185n3

complexity 6, 13, 16, 22, 32, 69, 78, 130–1, 159, 181, 194–6

concentration 8–17, 19, 22, 24–8, 32–3, 55, 57, 60, 63, 75, 78, 80, 81, 132, 141, 150, 156, 163, 180, 184, 186, 192

concentric model 20, 33

concentric rings model 15, 19, 20

Concessionária Porto Novo 154

conflict 49, 64, 137–8, 141, 150, 157, 181, 194, 196

connectivity 11, 33, 160

Consortium of Urban Intervention 154

construction intensity coefficients 43

consumption 1, 5, 9, 33, 49, 51, 57, 60, 143, 161–3, 176–81, 195

contradictions 18, 43, 49, 57, 73, 93, 96, 130–1, 163, 176, 186–7, 197

Corporación Antiguo Puerto Madero 132, 135

corporate business 65, 130

corporations 6, 9, 11, 74, 93, 102, 131, 138, 141, 143–4, 146, 159, 178, 181–4, 195–7

cosmopolitanism 8

cost-benefit 33

creative destruction 73, 88, 161, 165, 194

credit default swaps 72–3, 76–82

credit instruments 35, 165

credit scoring 92

cross-border flows 66; cross-border markets 67

customer profiling 92

debt 1–2, 35, 37, 73–8, 80, 82–7, 95, 97, 99, 103, 114, 118, 123–4, 136, 144, 159, 165, 168, 171–2, 182, 184, 196; debt-based 1; debt flows 11; debt instruments 84, 144, 165, 171–2; debt obligations 5, 38, 69, 72, 73, 76, 164; debt positions 5

decentralization 17, 22, 33, 48, 86

demand-supply analysis 30
density 8, 29–30, 32, 40, 42, *47*, 46–8, 58, 88, 131–3, 155–6, 179, 193
deregulation 2, 11, 30, 37–9, 61, 105, 124, 135, 159, 166, 170, 189
derivatives 2, 61, 73–5, 79–81, 89n8, 159–60, 167, 172, 194
design flaw, design 3, 12, 23, 30, *47*, 79, 104, 132, 135–43, 152, 159, 161, 163, 178, 192
devices of globalization 2, 64
differential income 47
differential rent 55
differentiation 7–8, 16–18, 49, 51, 54, 61n1, 181
discontinuities 33, 43
diseconomies 9, 189
disequilibrium 33, 186
dispersal 25
Distrito Criativo do Porto 152

economic cycles 5, 164, 195
economic growth 31, 73, 127, 149, 165, 187
economic modeling 13, 33
economic rationality 11, 37, 163
economic savings investments 61
economies of scale 10, 14, 28, 32, 53, 66, 119, 182
emerging economies 77, 93
endorsable mortgage mutual 98
energy and infrastructure investment trusts (FIBRA-E) 169; energy and infrastructure 174
enhancement charges 48
Enrique Peña Nieto 109
entertainment / theme centers 177
environment 3, 9, 11, 16–17, 40, 60, 65, 78, 83, 92, 102–3, 133, 145, 160, 163, 165, 176, 183, 188, 192, 197
equilibrium 10, 12–16, 20, 27–9, 33–4, 41, 43, 60, 76, 137, 187
equity 2, 3, 25, 42, 55, 68, 71, 73, 77, 84, 86–7, 89n19, 97, 115, 167, 171–2, 179; equity funds 65, 163, 167, 171, 182–3; equity loans 71, 74, 122; equity markets 61, 84, 87
equity debts REITs 84 *see also* equity
equity funds *see* equity
Exchange Traded Funds 85
expropriation 48
extension 17, 40, 142, 157n1, 177
externalities 27, 30, 34, 84, 86, 131

Fannie Mae 67–8, 105
Faria Lima 132
fashion / specialty center 177; fashion 37, 48, 138, 180
Federal Deposit Insurance Corporation 85
Federal Savings Bank 118
FIBRAS 169–71, 173–4, 185n5, 195
fictitious capital 35
finance capital 55, 66
financial assets 1, 7, 10, 33, 35–6, 52, 61n1, 64, 74, 84, 146, 154, 159, 168, 172, 194, 197
financial capital 2, 5–6, 66, 72, 116, 194, 198
financial crisis 72, 76, 78, 81–2
Financial Crisis Inquiry Commission *see* financial crisis
financial deregulation 2, 61, 170
financial derivative market 88
financial expropriation 66
financial innovations 37, 170
financial institutions 1, 26, 31, 66, 68, 70, 73, 78, 80–3, 98, 105, 109, 111, 113, 161, 171, 192
financial instruments 2, 33, 63, 159, 163, *168*, 170, 176, 180; financing infrastructure 176; housing industry 3; investment funds 5, 64, 159, 164, 166–7, 169, 171, 180, 198; liquidity in a real estate market 3; public-private company 154; real estate development 4, 10, 55, 138–9, 155–6, 193
financial market 86
financial rationale 10, 63, 97, 103–4, 125, 163, 186, 188
financial weapons of mass destruction 79, 160
Fiscal Arrangement Act 168
Fitch Ratings 76, 86
fixed assets 67, 88
flows of capital 179, 186
Fondo Nacional de Garantías 121
Fondo Solidario para la Vivienda 99
foreclosure 6, 38, 69, 72, 77, 80, 89n8, 89n13, 92, 95, 105, 115, 123–4, 127, 187, 188, 190–1
formal economy 7, 52
forward contract 67
FOVI 105
FOVISSSTE 94, 104, *108*, 111, 129n2, 189
free dwellings program, free housing units' program 124

Freddie Mac 67–8, 71
functional specialization 10, 15, 27
Fundo de Garantia de Tempo de Serviço
    114, 118, 154
future contracts 67
future rental income 37
futures 1–2, 36; futures markets 7, 61n1

gentrification 12, 30–1, 40, 52, 55, 93,
    131–2, 137, 148, 151, 156, 163, 191–2,
    197–8
gentrified historic districts 6
geographic specialization 12, 33
geographical economics 7
ghost town 28
Ginnie Mae 67, 105
global city 64; network 2–3, 9, 11, 15,
    25–6, 41, 52, 56, 63–5, 67, 80, 86, 96,
    121, 132, 142, 149, 150–1, 161, 171,
    180, 192–3
global financial network 2–3
global scale 52, 180, 195
government bonds 35, 83–4, 167
Government Sponsored Entities 67, 70
Great Recession 73, 77, 81–2, 89n7, 89n8,
    91, 116, 191
Greenspan, Allan 71, 75, 80–1
gross domestic product (GDP) 73, 81, 87,
    94, 98–9, *115*–17, 119–20, *120*, 122,
    127, 145, 160–1, 164, 166, 179, 180,
    190, 195
ground rent 8, 35, 37
Grundrisse 9

Harvey, David 6, 35, 37–8, 54, 80, 137,
    160, 194, 197
hedge 59, 67, 68, 70, 72, 80, 82–5, 87,
    109, 159, 167, 189
hedge funds 5, 10, 36, 75, 80–1, 83, 88,
    124, 171
hedging 61, 66, 68, 70–1, 81–2, 85, 105,
    122, 136
hexagon marketing principle 48, *50*
hexagonal structures 15, 34
high-density areas, high density 47, 179;
    high-density buildings 30
high-income 21–4, 51, 125, 135, 150,
    178
high-risk 77, 81, 92–3
investment practices 2
highest and best use value 84
hinterland 14, 17, 20, 32, 149
historic centers 27, 52, 56, 130

historic preservation 52, 58
HITO platform 111
home equity loans 71, 74, 122
housing 3, 7, 9, 11–3, 17, 20, 24–6, 28–9,
    31–2, 38, 43–6, *44*, 52, 63–6, 91, 93–4,
    97–9, 101, 104, 111, 116, 118–23, 125,
    132, 138–9, 154, 161–2, 188, 190–1,
    197; affordable 29, 104, 143, 146,
    188–9; housing policies 92, 101, 121,
    125, 188; housing production 94, 99,
    104, 118–19, 189; housing program 3,
    93, 95, 96, 189; housing projects 47,
    57, 99, 100; housing stock 3, 38, 53,
    92–3, 96, 99, 111, 117–108, 129n1, 149,
    189; housing developments 6, 12, 21,
    23, 28, 30–1, 41, 54, 56, 63, 72, 96, 98,
    101–2, 109, 111, 116–17, 122, 124, 132,
    137, 149, 156, 157n5, 189, 192; social
    housing units, housing units 3, 27–9,
    65, 69, 92, 96, 101, 104, 116–17, 122–4,
    139, 152, 188–90
Housing and Urbanization Services 99
housing bubbles 28, 31, 38, 41, 77, 81–2,
    92, 113, 116, 119, 124–5, 188
housing corporations 102
housing deficit 69, 96, 97, 116–17,
    119–20, 122, 189
housing financial system 114, 118
housing indicators system (SNIIV),
    Housing Registration System (RUV)
    113
Housing Operation and Banking Finance
    Fund 104
housing policy 120; *see also* housing
housing rental schemes 52
housing schemes 119
Hoyt, Homer 21–2
Hybrid REITs 84
hyper-densification 42, 56

illicit activities 10
imbalances 56, 59, 161, 175
incentives 1, 28–9, 64, 98, 109, 130, 148,
    157, 167, 178
inclusionary housing ordinances 3
income polarization 22, 31–2
indexation 95, 109, 122–3
industrial location 12–13
Industrial REITs 87
inefficiencies 25, 27, 32, 34, 48, 59, 96–7,
    109, 118, 155
inequalities 60, 64–5, 73, 96, 131–2, 141,
    165, 183

INFONAVIT 94, 102, *103*, 104, *108*, 109, 111, *113–14*, 129n2, 189
informal 7, 25, 28, 32, 49, 50, 52–4, 62n4, 94, 101, 109, 116, 118, 120, 122, 139, 141, 143, 145, 151–2, 157n5, 177–8, 188, 190, 193; economy 50, 94; housing 28, 53, 101, 157n5
infrastructure 3, 5, 10, 11, 25, 28–32, 36, 38–42, 47, 53–4, 56–60, 63–5, 93, 102, 104, 109, 117, 122, 130–9, 142–4, 146, 148, 151–6, 159–61, 163–76, 181, 185n7, 191–8; enterprises 10, 11, 36–7, 60, 93, 104, 109, 155, 166, 173; FIBRA E 174–5; public facilities 53, 117, 138; public-private 47, 56–7, 131, 135–6, 138, 141, 143–4, 154–5, 157, 163, 166–7, 169, 186, 188, 191; projects 11, 31, 36, 38, 40, 58, 63, 65, 93, 130, 142, 160, 162, 164–8, 170, 172, 174–5, 181, 195
infrastructure and real estate trusts 170
Initial Public Offerings (IPOs) 115
instability 1, 5, 51, 102, 124–5
institutional framework 55, 92
Instituto de Crédito Territorial 122
Instituto Nacional de Vivienda Social y Reforma Urbana (INURBE) 120, 122
insurance companies 5, 10, 64–5, 70, 76, 95, 109, 124–5, 127, 159, 164, 169, 171, 174, 178, 191, 195
intensity of use 12
Inter-American Development Bank 10, 164, 166, 175
intermediate circuit 45, 49–50, 52
International Bank for Reconstruction and Development 60
International Council of Shopping Centers 176
International Development Bank 94, 126, 164, 169, 175
International Financial Corporation 105
International Monetary Fund 66, 69, 81, 104, 114, 169
interurban mobility 51
Investment Project Trust Certificates, investment project 2, 169, 171
ISSSTE 111, 170, 174

Jacobs, Jane 32
Jaramillo, Samuel 8, 40, 41, 46, 55, 123

labor 5, 8, 9, 12, 13, 15–16, 18, 32–3, 35, 37, 41, 53, 125, 136, 162

land 2, 4–7, 10, 12–13, 21, 25, 28, 32, 34, 36–41; market 37, 40, 42, 47, 57–9, 96, 102, 131, 156, 176; price 2, 6, 13, 20, 27–31, 33–4, 39–43, *44–47*, 48, 53–5, 57, 102, 120, 123, 143, 152, 178, 197; readjustments 46; taxes 47, 57; use 6, 10, 12–13, 17–23, 27, 29–30, 33–4, 40–1, 43, 45, 48–9, 55, 57–60, 65, 84, 100, 132–3, 137, 139, 148, 161, 179, 186; value 3, 6, 12, 17, *19–21*, 28–9, 33–4, 37, 40–1, 46, 54, 57, 130, 132–3, 136–7, 141, 148, 151–2, 156, 178–80, 188, 191–2, 194–6, 198
large-scale urban operations 3, 57
Latin America 3, 38–9, 55, 58, 60, 72, 88, 97, 125, 127, 130, 132, 162, 166, 175, 177–8, 180, 188–9, 195
leasing 98–99, 115, 117, 173, 182, 196
Lefebvre, Henri 8, 9, 11, 54, 63, 160
Letras de Cambio 98
Letras de Crédito Hipotecario 94
Letras de Crédito Imobiliário (LCI) 118
leverage 2, 37, 65, 75–6, 78, 80–2, 99, 116, 119, 124, 159, 161, 169, 171, 184, 188, 194, 196
liabilities 70, 74–6, 94, 168, 172, 178, 182
Lipietz, Alain 8, 57, 117
liquid assets 84, 88
location theory 13, 34
locational equilibrium 27, 34
long-term investment 38, 86, 126, 131, 146, 160, 164, 188, 191
long-term structural variations *44*
Loop 17, 20
Lösch, August 15, 49
low-income 20, 26, 28, 32, 53, 58, 63, 72, 93, 98, 100–2, 111, 117, 119–22, 145, 149, 151–2, 188–90, 192–3; citizens 20; housing 63, 72, 98, 101, 111; households 100, 120, 121; inner-city residents 26; population 28, 32, 53, 93, 102, 122, 152, 188, 193; residents 58, 145
lower circuit 49–50, 52–3
Lula da Silva 116

malls 3, 86, 132, 138, 177–80
marginal cost 43
market bubbles 40, 86; equilibrium 10, 12–16, 20, 27–9, 33–4, 41, 43, 60, 76, 137, 187; forces 6, 49, 56, 60, 66; segmentation 48
Marx, Karl 2, 8, 9, 35–9, 159

Marxist 2, 36, 42, 66, 155, 165
megaprojects 3, 130–1, 138, 141, 181, 191
metabolism 18
Mexican miracle 104
Mexican Stock Exchange Market (BMV) 173
Mexico 1, 3, 69, 93–5, 102–3, 105, 109, 111, 117, 126–7, 129n1, 129n7, 138, 142, 146, 164, 167, 169–71, 173, 175, 177, 179–80, 181, 188–90, 195–6
Mexico City 6, 138, 192
Mi casa ya housing program 125
Minha Casa, Minha Vida 93, 113, 116, 190
Ministerio de Urbanismo y Vivienda MINUVI 96, 99, *100*
Minsky, Hyman 61n2, 67, 85, 124, 161
money laundering 3, 10, 74, 88, 145, 149, 157, 161, 193–4
monopoly price 51, 54
Moody's 76, 86, *108*
mortgage backed securities 1–2, 5, 67–9, 71–2, 76, 88n6, 89n8, 89n9, 94, 105, 114, 127, 163, 188, 198
mortgages 2, 9, 27, 35–7, 52, 61, 64, 67–78, 83–4, 87, 89n7, 89n9, 89n19, 91–9, 102–5, 109, 111, 114, 116, 118–19, 124–5, 127, 146, 148, 161, 181, 188–90, 193, 198; bonds 35, 61, 67, 74, 81, 83–4, 94–5, 98, 126–7, 154–5, 165–7, 169, 170, 190, 193–4; REITs 84, 87; securitization 65, 71, 94, 119, 126–7; securitizer 105
multi-nuclear model 34
multinational corporations 6, 93, 141
municipally-issued bonds (CEPACs) 154, 193
mutual funds 60, 64, 70, 83, 85, 98, 165, 167

NAREIT Equity Index 86–7, 89n19
NASDAQ 179
National Association of Convenience and Department Stores (ANTAD) 179
National Registry of Territorial Reserves (RENARET) 113
National Banking and Securities Commission (CNBV) 172
National Commission of the Retirement Savings System (CONSAR) 169–71
National Infrastructure Fund (FONADIN) 172

National Institute of Social Housing and urban reform 122
neighborhood center 177, 179–80
Neighborhood Recovery Program 101
neoclassical economic paradigm 26
neoclassical theories 55
neoliberal city 49, 175
New York Stock Exchange 83, 179
NINJA loans 35, 78
North American Free Trade Agreement 142

O'Sullivan, Arthur 27
Odebrecht 149
Office REITs 86
open-air strip centers 177
optimal allocation 57
options 12, 25, 32, 67, 78, 88n2, 146, 181, 187; public policy options 3
Orçamento Geral da União 117
outlet centers 86, 177
over-accumulation 11, 36–7, 55, 64, 92, 160
over-the-counter 81

Panama Canal 145, 149, 193
Panama City 144, *147*, 149, 182, 192–3
passive speculation *46*
patterns 5–6, 13, 15, 19, 21, 23, 24, 31–3, 58–9, 63–4, 83–4, 92, 103, 156, 160, 176
pension funds 5, 10, 61, 64–5, 95, 109, 115, 118, 124–7, 159, 164, 169–72, 174, 178–9, 191, 195–6
Pension Funds Investment Societies 170
Polany, Karl 39
polarization 22, 31–3, 49, 51, 64, 93, 103, 135, 156, 175, 183, 187, 191
Polygons of Action 139
polynuclear model *22, 23*
Porter, Michael 26
Porto Maravilha 132, 151–7, *153*, 193
power centers 86, 177
predatory lending 92, 188
predictive behavior 92
premiums 78
private interest 33
privatization 11, 60–1, 93, *93*, 97, 104, 114, 122, 135, 181, 184, 189
production costs 10, 51, 100
productive activities 8, 11–12, 15, 35, 146, 163, 166, 175
progressive housing program 96

public facilities 53, 117, 138
public infrastructure projects 58, 130, 165, 174
public interest 29, 34, 48, 84, 93, 131–3, 143, 176, 187–8, 192
public interventions 48, 141
public policies 93, 97, 122, 124, 150, 152, 160, 163, 198
public spaces 26, 29, 42, 100, 101, 132, 137, 142, 151, 154, 156, 192
public transport systems 25, 34, 86, 133
public-private partnerships 56–7, 131, 144, 157, 163, 166–7, 169, 191
Puerto Madero 132–7, *134*, 155, 192–3
Punta Pacífica 132, 144, 146–7, 149, 157n4, 192–3

radial sectors model *21*, 33
rational patterns 59
real estate 1–11, 21, 24–5, 27–31, 34–5, 37–8, 40, 42–3, 45, 47, 49–56, 58, 60, 63–6, 68–70, 73, 83–8, 89n19, 90n19, 90n20, 91–2, 95, 97, 99–102, 111, 113–14, 116–19, 122–5, 129n6, 131–3, 136–9, 141–6, 148, 150–2, 154–7, 160–2, 164, 166, 169–82, 184, 185n2, 185n5, 186–98
Real Estate Financial System (SFI) 118
real estate investment trusts 1–2, 5, 37, 66, 70, 83–4, 166, 173, 182, 184, 195
real estate mortgage investment conduits 68
real estate tourism 182
real estate transfer tax 173
real estate trust certificates (CBFIs) 173
redistributive 48, 57, 117, 138, 150
refinancing 92, 188
Região Portuária REIT 154
regional center 177
REITs 2, 5, 70, 83–7, 89n19, 90n21, 144, 163, 169, 171, 173, 178, 182–4, 185n3, 185n9, 194–6; financial instruments 2, 33, 63, 159, 163, *168*, 170, 176, 180; mortgage REITs 84, 87; mortgage-backed securities 1–2, 5, 67–9, 71–2, 76, 88n6, 89n8, 89n9, 94, 105, 114, 127, 163, 188, 198; pension funds *see* pension funds; private equity funds 65, 163, 167, 171, 182–3; private investment 3, 70, 131–2, 135, 137, 143, 150, 164, 171, 192
Residential Leasing Fund 117

residential mortgage-backed securities 68, 71, 88n6
residential REITs 86
residential stability 51
restrictions 3, 6, 13, 29, 43, 58, 77, 81, 81, 98, 150, 157, 185n7, 190, 194
retail centers 176, 178–9
retail REITs 86–7; Big- box retail spaces 86; community centers 86, 180; Super-regional malls 86; Shopping centers 3, 17, 22, 27–8, 30–1, 40, 63, 65, 86, 135, 161–2, 173, 176–9, 195
revenues 33, 35–6, 114, 124, 136–9, 143–4, 150, 154–7, 159, 163, 165, 172–4, 179, 181–3, 196
Rio de Janeiro 22, 132, 151–4, *153*, 162, 193
Riviera Maya 181

Santa Fe 132, 138–41, *140*, 143–4, 192
Santiago de Chile 202, 206, 210
Santos, Milton 5, 43, 48–9, *50*, 52
Sao Paulo 22
saturation 56, 60, 156
Saving and Loans Crisis, e US savings and loan systems 85
scarce resources 53, 57, 59
scarcity mechanisms 48
second circuit of capital accumulation 11, 160
second-lien mortgages 71
second-tier development Banks, second-tier banks 105, 119, 129
secondary market 67–8, 70, 92, 94, 105, 118, 126, 155, 187
securities 1–2, 5, 36–8, 61, 67–72, 75–84, 88n5, 88n6, 89n7, 89n8, 89n9, 91–2, 94–5, 105, 109, 111, 114, 118, 119, 125–7, 133, 144, 155, 159–60, 163, 167–8, 170, 172–4, 178, 188, 191, 198
Securities and Exchange Commission 77–8, 83, 155
Segregation 6, 8, 16–18, 27, 31–2, 42, 47, 49, 54, 56, 64–5, 88, 103, 121–2, 124–5, 131, 135, 137, 152, 156, 158n6, 178, 184, 187, 189–91, 193, 197
self-fulfilling prophecies 103
self-help housing 118
self-regulating market 39
self-storage REITs 87
Servicios Metropolitanos 141
shadow economy 52, 196
shares 35–6, 87–8, 166

shell companies 74, 146, 149, 151
shopping centers 3, 22, 27–8, 30–1, 40, 63, 65, 86, 135, 161–2, 173, 176–9, 195
Simmel, George 8
Sistema Brasileiro de Poupança e Empréstimo 114
Sistema de Financiamento Habitacional, Sistema Financeiro de Habitação 114, 118
size 8, 10, 13–14, 32, 59, 74, 77, 86, *107*, 122, 155, 164, 173, 177
social cohesion 39
social configurations 57
Social Development Fund 117
social differentiation 54
social housing developments 12, 21, 28, 41, 54, 63, 98, 102, 132, 137, 156
Social Real Estate Management Entities 99
social segregation 18, 32, 103, 135, 137, 152, 156, 158n6, 178, 189
social system 49
socially owned land 102
Sociedad Hipotecaria Federal 69, 105, 111, 127
socio-spatial segregation 6, 17, 47, 56
sociotechnical devices 163
SOFOLES 105, 109, 111, 127
SOFOMES 105, 109, 111
spatial competition 12, 17
spatial dialectics 5
spatial differentiation 49
spatial divisions 57
spatial economic circuits 5, 50
spatial economics 12, 33, 59
spatial economy 6, 22, 45, 48–9, 52, 58, 60
spatial equilibrium 15, 29, 34, 187
spatial fix 11, 55, 63, 88, 113, 149, 160, 186, 191, 194
spatial fragmentation 18, 22, 32, 62, 103, 135, 137, 141, 152, 156, 184, 186, 189
spatial order 57
spatial rationality 12, 33
spatial solidity 182
spatial structure 2, 7, 15, 17, 26, 33, 35, 42–3, 45, 47, 49–52, 54, 58–9, 63, 65–6
spatial-temporal fixes 66
spatial-temporal structure 49
Special Controlled Development Zone 139
special purpose corporation 74
special purpose entities 74
special purpose vehicles 74, 105, 126

specialties REITs 87; gas stations 87; Golf courses 87, 185n8; movie theaters 87; prisons 87
speculation 6, 27, 37–8, 42–3, 46, 54, 73, 88, 137, 143, 155, 175, 194, 197
speculative capital 55
speculative demand 41, 55
spreads 64, 78, 92
Standard and Poor's 76
State intervention 41, 47–8, 56–7
State-owned enterprises 60, 104
stock market 28, 36, 51, 60–1, 64, 83–4, 109, 116, 165, 168, 172–3, 181
stocks 36–7, 67, 83–4, 103, 105, 122, 159, 188; capital values 37, 59; RMBS stocks 105
stratification 49, 139
structural coherence 11, 33
structural deficit 53, 125
structural finance 61
structural reforms 60, 66, 93, 104, 156
structured finance 67, 69, 71, 74–5, 77–8, 126, 131, 144, 148
subprime 38, 59, 65, 67, 69–72, 77, 79–80, 85, 91–3, 98, 105, 109, 114, 121, 123, 170, 179, 183–4, 189, 196
subsidies 1, 10, 52, 96–101, 104, 109, 116, 117, 119, 120, 122–3, 125, 131, 187, 190, 198
succession 16
superregional center 177
supply and demand 6, 30, 42, 53–4, 59, 86
surplus capital 11, 37, 63
surplus value 35, 37, 39, 43, 46, 65, 92, 160
Swaps 1–2, 72–3, 76–80, 82
synthetic CDOs 73, 75–7, 82
System of Housing Subsidies 101

Tax Increment Financing (TIF) 155, 194
tax revenues 36, 136, 138
Tecnologías Urbanas de Barcelona 135
territorial balance 3, 55, 192
territorial base 49
territorial dispersion 10
territorial tax collection 48
territorialization 8
territory 5, 6, 8, 17, 33–4, 46, 56, 63, 86, 93, 102, 130, 139, 156, 160–1, 165, 175–6, 180–1, 186–8, 191–2, 194–5
theories of land rent 2, 39
three circuits 48–9, *50–1*
thresholds 14

Titularizadora Colombiana 95, 127, 129n9
Topalov, Christian 8–9, 43, 117
tourism 3, 9, 30, 41, 55, 58, 146, 152, 156, 180–4, 193, 196
tourist enclaves 3, 183, 196
tourist resorts 3, 130
trade 12, 36, 53, 64, 66–7, 69–70, 83, 86, 91, 93, 142, 144–6, 149, 151, 163, 170, 173, 180, 183, 188, 193, 197
tranched index 79
tranches 72–7, 79, 81, 88n5, 89n9, 111
transaction costs 61, 84
transportation costs 7, 12, 13, 33–4, 176, 195
triple-A securities, AAA securities 71, 76
trusts 2, 5, 70, 83–5, 166, 169–70, 172–74, 178, 182, 184, 185n5, 195

US Treasury 66
Ullman, Edward 14, 22–5
Unidades de Medida y Actualización 109
unit investment trusts 85
unregulated financial instruments 59
upper circuit 49–51, 53, 60
urban agglomerations 10, 58
urban concentration 8, 13, 32
urban development 2, 8, 42, 55, 109, 131, 155–6, 163, 182, 187, 192–3; amenities and infrastructure 155; development strategies 109, 163, 180; financial capital 192; *see also* financial capital; increasing land values 179; in Latin America 1, 130; socio-spatial segregation 6, 17, 47, 56; *see also* segregation; shareholders 83, 92, 172, 181
urban development program 139
urban ecology 15, 33
urban economic models 31
urban economy 2–3, 5, 10–11, 29, 48–9, 50, 163
urban expansion 18, 25, 41, 45, 55, 58, 98, 179
urban land rent theories 39

urban nodes 64
urban operation 1–3, 28, 31, 41–2, 48, 54, 57–8, 65, 130, 132, 135–7, 154–7, 158n9
urban planning instruments 5, 9, 33, 191, 194; large capital investments 27–8, 34; central power groups 9
urban projects 31, 51, 65, 130–1, 133, 137–9, 144, 150, 152, 155–6, 163, 175, 197
urban regeneration 31, 47, 101
urban renewal 51, 54
urban renovation 58
urban transport 7, 24–5, 31
US Federal Reserve 68, 71, 77, 80

vacant housing developments 6, 30–1; gentrification 12, 30–1, 40, 52, 55, 93, 131–2, 137, 148, 151, 156, 163, 191–2, 197–8; in Mexico City 6; overproduction 27, 30, 54, 148, 197; spatial segregation 6, 8, 17, 31, 47, 54, 56; Value-Added Tax, added value 39, 176
value-capture instruments 3; financing infrastructure 176; public policy 3, 6–7, 17, 29, 31, 33, 34, 54, 93, 122, 186
Von Thünen 13

Washington Consensus 60, 66, 97, 104
Weber, Alfred 7, 12, 13, 155–6
World Bank 1, 60, 93, 97, 104–5, 114, 126, 129n3, 169, 185n6, 197
World Trade Organization 183
World Urban Forum 151

zoning 8, 17, 29, 43, 44, 48, 57, 84, 86, 90n20, 130–1, 137, 139, 148, 152, 155, 158n9; density regulations 48, 156; ordinances 3, 29, 85, 137; provisions 10, 17, 40–1, 43, 57, 69, 98, 104, 111, 117–18, 148, 152, 156, 158n9, 161, 169, 172; restrictions 3, 6, 13, 29, 41, 43, 48, 77, 81, 91, 98, 150, 157, 185n7, 190, 194

Printed in the United States
by Baker & Taylor Publisher Services